Multi-Objective Optimization of Industrial Power Generation Systems:

Emerging Research and Opportunities

Timothy Ganesan
Royal Bank of Canada, Canada

A volume in the Advances in Civil and Industrial Engineering (ACIE) Book Series

IGI Global
DISSEMINATOR OF KNOWLEDGE

Published in the United States of America by
IGI Global
Engineering Science Reference (an imprint of IGI Global)
701 E. Chocolate Avenue
Hershey PA, USA 17033
Tel: 717-533-8845
Fax: 717-533-8661
E-mail: cust@igi-global.com
Web site: http://www.igi-global.com

Library of Congress Cataloging-in-Publication Data

Names: Ganesan, Timothy, author.
Title: Multi-objective optimization of industrial power generation systems : emerging research and opportunities / by Timothy Ganesan.
Description: Hershey, PA : Engineering Science Reference, 2020. | Includes bibliographical references and index. | Summary: "This book describes and discusses some key applications of CI frameworks in power generation systems"-- Provided by publisher.
Identifiers: LCCN 2019032968 (print) | LCCN 2019032969 (ebook) | ISBN 9781799817109 (h/c) | ISBN 9781799817116 (s/c) | ISBN 9781799817123 (ebook)
Subjects: LCSH: Electric power systems--Control--Data processing. | Artificial intelligence. | Multiple criteria decision making.
Classification: LCC TK1007 .G36 2020 (print) | LCC TK1007 (ebook) | DDC 621.31/21028563--dc23
LC record available at https://lccn.loc.gov/2019032968
LC ebook record available at https://lccn.loc.gov/2019032969

This book is published in the IGI Global book series Advances in Civil and Industrial Engineering (ACIE) (ISSN: 2326-6139; eISSN: 2326-6155)

British Cataloguing in Publication Data
A Cataloguing in Publication record for this book is available from the British Library.

Advances in Civil and Industrial Engineering (ACIE) Book Series

ISSN:2326-6139
EISSN:2326-6155

Editor-in-Chief: Ioan Constantin Dima, University Valahia of Târgovişte, Romania

MISSION

Private and public sector infrastructures begin to age, or require change in the face of developing technologies, the fields of civil and industrial engineering have become increasingly important as a method to mitigate and manage these changes. As governments and the public at large begin to grapple with climate change and growing populations, civil engineering has become more interdisciplinary and the need for publications that discuss the rapid changes and advancements in the field have become more in-demand. Additionally, private corporations and companies are facing similar changes and challenges, with the pressure for new and innovative methods being placed on those involved in industrial engineering.

The **Advances in Civil and Industrial Engineering (ACIE) Book Series** aims to present research and methodology that will provide solutions and discussions to meet such needs. The latest methodologies, applications, tools, and analysis will be published through the books included in **ACIE** in order to keep the available research in civil and industrial engineering as current and timely as possible.

COVERAGE

- Production Planning and Control
- Ergonomics
- Transportation Engineering
- Earthquake engineering
- Structural Engineering
- Productivity
- Engineering Economics
- Construction Engineering
- Materials Management
- Operations Research

IGI Global is currently accepting manuscripts for publication within this series. To submit a proposal for a volume in this series, please contact our Acquisition Editors at Acquisitions@igi-global.com or visit: http://www.igi-global.com/publish/.

The Advances in Civil and Industrial Engineering (ACIE) Book Series (ISSN 2326-6139) is published by IGI Global, 701 E. Chocolate Avenue, Hershey, PA 17033-1240, USA, www.igi-global.com. This series is composed of titles available for purchase individually; each title is edited to be contextually exclusive from any other title within the series. For pricing and ordering information please visit http://www.igi-global.com/book-series/advances-civil-industrial-engineering/73673. Postmaster: Send all address changes to above address.

Titles in this Series

For a list of additional titles in this series, please visit:
https://www.igi-global.com/book-series/advances-civil-industrial-engineering/73673

Re-Coding Homes Through Flexible Interiors Emerging Research and Opportunities
Nilüfer Saglar Onay (Istanbul Technical University, Turkey) S. Banu Garip (Istanbul Technical University, Turkey) and Ervin Garip (Istanbul Technical University, Turkey)
Engineering Science Reference • © 2020 • 165pp • H/C (ISBN: 9781522589587) • US $175.00 (our price)

Green Building Management and Smart Automation
Arun Solanki (Gautam Buddha University, India) and Anand Nayyar (Duy Tan University, Vietnam)
Engineering Science Reference • © 2020 • 312pp • H/C (ISBN: 9781522597544) • US $215.00 (our price)

Handbook of Research on Implementation and Deployment of IoT Projects in Smart Cities
Krishnan Saravanan (Anna University Chennai – Regional Office Tirunelveli, India) Golden Julie (Anna University, India) and Harold Robinson (SCAD College of Engineering and Technology, India)
Engineering Science Reference • © 2019 • 415pp • H/C (ISBN: 9781522591993) • US $295.00 (our price)

Handbook of Research on Digital Research Methods and Architectural Tools in Urban Planning and Design
Hisham Abusaada (Housing and Building National Research Center, Egypt) Carsten Vellguth (German Academic Exchange Service, Germany) and Abeer Elshater (Ain Shams University, Egypt)
Engineering Science Reference • © 2019 • 445pp • H/C (ISBN: 9781522592389) • US $265.00 (our price)

IGI Global
DISSEMINATOR OF KNOWLEDGE

701 East Chocolate Avenue, Hershey, PA 17033, USA
Tel: 717-533-8845 x100 • Fax: 717-533-8661
E-Mail: cust@igi-global.com • www.igi-global.com

Table of Contents

Foreword

The primary aim of this book, *Multi-Objective Optimization of Industrial Power Generation Systems: Emerging Research and Opportunities* is to provide insights on the implementation of current multi-objective (MO) optimization techniques to application problems in power engineering. This book provides recent developments of computational intelligence in the power generation industry covering bioenergy, solar power, distributed generation, coal power, hydro and hydrothermal systems, gas turbine plants, combined cycle systems and nuclear plants. In addition, the book primarily focuses on complex MO optimization problems involving: solar power technologies, industrial gas turbine systems and biofuel supply chains. This book is invaluable for readers serving as a guideline to conduct research and tackle real-world industrial MO problems in the energy industry. It provides algorithmic details on current state-of-the-art MO optimization techniques applied in the power industry - the results of these implementations are also critically analyzed and discussed throughout the book.

This book blazes the trail for emerging research areas in MO techniques - e.g. evolutionary algorithms, swarm intelligence as well as symbolic programming paradigms (analytical and genetic programming). The author introduces fundamental elements, computational procedures and detail industrial applications in MO optimization. These factors provide the essential tools and 'know-hows' for building and applying efficient MO algorithms. In power generation, practical utilization of MO techniques has become of utmost importance providing various advantages to organization and decision makers globally. Among these advantages are; timely delivery of service, adherence to service level agreements (SLA), optimization of existing systems, improved utilization of resources and efficient system management.

Another key contribution of this book is the novel ideas presented on improving existing methodologies in MO optimization. For instance, in

Chapter 2, a Type-2 Fuzzy Logic approach was utilized in the MO solar irrigation problem (which was plagued by uncertainties). In the design of alternative energy systems such as solar technologies, the engineer/decision maker would often encounter noise sources when their system interacts with the environment (e.g. solar insolation and ambient temperature fluctuations). The author presents a novel idea where a generalized problem is constructed and solved within a Type-2 Fuzzy formulation.

In Chapter 3, a chaos-based mechanism was employed to enhance the differential evolution computational technique. By manipulating the chaos levels of the solution method, the optimization efforts was seen to significantly improve. The interesting concept of stochastic engines is introduced and employed in Chapter 4 for tackling the gas turbine waste heat recovery MO optimization problem. There the author shows that each stochastic engine uniquely influences the optimization results - when applied to highly complex problems (MO problems with four or more objective functions). In that chapter, the idea of manipulating the factors in stochastic engines is given (probability distribution; shape, scale and location) with the idea of improving their effectiveness.

In Chapters 5 and 6, Lévy flight enhancements and random matrices were incorporated into the optimization strategies for solving the high-complexity biofuel supply chain problem. These novel frameworks inspire further research works and pushes the limits of conventional solution strategies in MO optimization.

This valuable volume consists of the following research goals, topics and techniques:

- It provides comprehensive guidelines for engineers/decision makers for building algorithmic solution methods for MO optimization of power generation systems.
- It explores various MO settings and scenarios in real-world industrial power generation systems.
- It discusses and critically analyzes the results obtained from the application of different types of solutions strategies - and the impacts of algorithmic enhancements on the optimization results (positive and negative impacts).
- It also supplies the reader with some of the behaviors and characteristics of the computational method during execution on these industrial problems.

- It provides the reader with potential research directions and ideas for constructing interesting hypotheses for further numerical experimentations.

MO applications using computational intelligence, specifically metaheuristics are currently growing at an unprecedented level. These applications range broadly across various industrial engineering sectors: supply chain management, system design, operations/maintenance, risk/reliability engineering, control systems, process modelling, energy optimization and resource utilization. Although this book focuses on three specific applications: solar power technologies, industrial gas turbine systems and biofuel supply chains, the solution strategies and the 'thinking framework' utilized in this book could be extended to other areas in industrial engineering.

The discussions provided by the author are rich and multifaceted covering various aspects of industrial applications as well as solution strategies. The variety of topics covered in this book is extracted from the author's vast experience in various research works in MO industrial optimization. The author combines two styles. The first is a 'data-driven' approach which emerges from a 'problem-based solution' philosophy frequently used in engineering and the applied sciences. The second one is a 'theoretical model-centric approach' springing from fundamentals in mathematics and physics - where he wields diverse mathematical tools from highly theoretical fields. This blend of styles is very suitable when dealing with complex problems as encountered in this book as well as problems where the lines between theory and application are blurred.

To the author of the book, Timothy Ganesan, we would like to extend our gratitude and appreciation for utilizing your analytical skills, expertise and research experience to generate this volume - which is invaluable for other researchers and decision makers in academia and industry alike. We are also thankful to the publisher IGI Global for providing an avenue for experts like the author to publish their current developments and emerging research methodologies - helping them increase the accessibility of their work to a larger scientific community.

To the reader, we wish that upon reading this book you will gain invaluable insights and useful knowledge - most importantly we wish that you obtain the knowledge to build effective techniques and tools for solving a vast array of problems in industrial optimization. Moreover, the content of this book will provide holistic and break through global framework and insight new knowledge for the international research scholars across the planet.

Sincerely,

Pandian Vasant
Universiti Teknologi PETRONAS, Malaysia

Preface

As economies become increasingly complex, so do their associated energy generation systems. Therefore, engineers and decision makers in this sector are spurred to seek out state-of-the-art approaches to deal with the rapid increase in system complexity. An effective strategy to deal with such scenarios is to employ methods in computational intelligence (CI). CI supplements the heuristics used by engineers thereby enhancing the cumulative analytic capacity towards effectively resolving complicated scenarios. CI could be split into two classes: predictive modeling and optimization. Predictive modeling deals with engineering problems involved in the development of effective models (or correlations) to describe specific systems. The higher the 'system complexity', the more sophisticated the predictive model. The notion of 'system complexity' employed here is defined as systems which have the following attributes:

- Multivariate
- Nonlinear and nonconvex
- Variables are highly correlated with one another
- Multiple constraints
- Variables riddled with various uncertainties
- Stochastic

Under such circumstances, more capable techniques for predictive modeling are required - since conventional regression approaches become ineffective. For instance, state-of-the-art techniques such as enhanced artificial neural nets (ANN) and support vector machine/regression (SVM/SVR) are more effective and relevant. Although both groups of techniques are very capable for handling a broad range of problems, support vector-based approaches are more suitable for problems involving data classification.

Besides predictive modeling and classification, problems have also become significantly complex in other areas of engineering such as system optimization. In those areas, a new class of CI techniques has slowly emerged in the last two decades; metaheuristics. These methods encompass a huge group of algorithmic techniques springing from two of the most dominant schools of thought; swarm and evolutionary computing. These optimization frameworks have been used broadly across the power industry. The objective of this book is to describe and discuss some key applications of CI frameworks in power generation systems. The motivation to write this book comes from the lack of scientific literature covering topics related to current implementation of CI in power generation. With this in mind, the first part of this book provides a comprehensive compilation of recent literature in power generation covering key major industries globally: bioenergy, solar power, distributed generation, coal power, hydro and hydrothermal systems, gas turbine plants, combined cycle systems and nuclear plants. In this section, these current developments are discussed in terms of description of the energy system, type of CI technique employed, effectiveness of the technique and detailed analyses of the CI application.

The remaining parts of this book contain CI applications to four real-world industrial systems in the power generation sector. The second chapter of this book describes the sizing and design optimization of a solar powered irrigation system. The optimization was carried out under two different considerations: deterministic and with uncertainty. Optimization of the solar powered irrigation system was mostly done using improved swarm intelligence and evolutionary programming. The fourth chapter of this book describes the optimization of a gas turbine (GT) system. In this system, the air inlet of the gas turbine was cooled using an absorption chiller (AC). This GT-AC system was optimized using modified evolutionary approaches. Chapter 5 deals with the optimization of a biomass fuel supply chain system. This supply chain was optimized using an array of enhanced swarm-based approaches: cuckoo search (CS), gravitational search algorithm (GSA) and particle swarm optimization (PSO). Note that although the cases employed in this book were based on industrial-scale applications, the data employed have been intentionally modified to respect corporate confidentiality.

Central insights on strategies in engineering optimization have been provided by many influential books, technical articles and internet resources.

Some of these resources appear in the reference section of this book. The interested reader may refer to these resources for further details regarding the optimization techniques, model formulation or numerical experiments. Alternatively, many of these details are also provided by various web resources.

Acknowledgment

I would like to thank Tenaga Nasional Berhad - Research (TNBR) for their support throughout the development of this book. Special thanks goes out to my research collaborators at Universiti Teknologi PETRONAS (UTP); Prof. Pandian Vasant and Prof. Irraivan Elamvazuthi as well as Prof. Ivan Zelinka from the Department of Computer Science, Faculty of Electrical Engineering and Computer Science, Technical University of Ostrava. I thank them for extending their resources and ideas; which has repeatedly come to my aid while writing this book. This book would not have come to fruition without the support of my close friends, colleagues, students and most importantly my family. I am very grateful to my fellow researchers and good friends at TNBR: Dr Shiraz Aris, Mr Ahmad Zulazlan Shah, Mr Hamdan Hassan, Dr Zainul Asri Mamat and Mr Saiful Adilin Shokri for their technical support and research insights which were critical for the material presented in this book. I would like to thank Ms Sima Alian, Mr Suren Lim Sinnadurai and Mr Jasvinder Singh Gill for their patience and continual support during the writing of this book. As always, I am particularly grateful to my loving and supportive family; my grandmother and aunts for always being there for me. I apologize again to my grandmother and aunts for sometimes neglecting my responsibilities at home due to my preoccupation with writing this book. To my editors and personnel at IGI Global as well as my reviewers; especially Miss Halle N. Frisco, thanks very much for finishing this journey with me.

Chapter 1
Computational Intelligence in Energy Generation

ABSTRACT

As economies become increasingly complex, so do their associated energy generation systems. Therefore, engineers and decision makers in this sector are spurred to seek out state-of-the-art approaches to deal with this rapid increase in system complexity. An effective strategy to deal with this scenario is to employ computational intelligence (CI) methods. CI supplements the heuristics used by the engineer—enhancing the cumulative analytic capacity to effectively resolve complicated scenarios. CI could be split to two classes: predictive modeling and optimization. In this chapter, past applications of CI in energy generation are discussed. The sectors presented here are renewable energy systems, distributed generation, nuclear power plants, coal power, and gas-fueled plants.

INTRODUCTION

Energy generation industries have become increasingly multifaceted in recent years. As economies become complex, so does the customer demands, regulatory requirements and environmental policies. Thus power suppliers have to reliably generate power while accommodating all these new constraints – imposed by the state as well as the customer. To function in such settings various aspects of the industry has evolved and become more complicated. The divisions affected by these new conditions are:

DOI: 10.4018/978-1-7998-1710-9.ch001

- Plant operations and management
- Supply chain planning and scheduling
- Power Transmission and Distribution
- Plant Design and Process Engineering

Therefore engineers and decision makers in this sector are spurred to seek out state-of-the-art approaches to deal with this rapid increase in system complexity. The global power industry has currently gone into the era of digitization - where the information and data-driven technologies are used to enhance energy transmission, distribution and generation (. These technologies become rapidly prevalent due to the availability of cheap sensors and powerful platforms to run data analytics. The abundant availability of cheap computational power is also a primary factor which contributes to the digitization of the electricity industry (Allegorico and Mantini, 2014; Yan 2016; Fu *et al.*, 2016; Simmhan *et al.*, 2013; Diamantoulakis *et al.*, 2015; Shyam *et al.*, 2015; Fang et a;., 2016). Digitization also plays an integral role in technologies involving renewable and alternative energy sources (Nabati and Thoben, 2017; Sheng., 2015; Viharos, *et al.*, 2013; Oró *et al.*, 2015). Figure 1 depicts an overview of energy systems which deliver electricity into the grid - where it is distributed and transmitted to the consumer:

Computational intelligence (CI) is an effective strategy that comes to aid when solving these problems (Verma *et al.*, 2016a; Paul *et al.*, 2013). CI supplements the heuristics used by the engineer enhancing the cumulative analytic capacity. This increases the rate of which these complex scenarios are effectively resolved. CI is usually targeted towards applications such as: predictive modeling and optimization. In both instances, these strategies become indispensable when the problem at hand portrays high-levels of complexity. For instance problems which are highly nonlinear, non-convex, contains a large number of variables/target objectives or has variables which are uncertain (or stochastic).

These problems if remain unsolved could cause major losses to industries in various energy sectors. Besides that, financial losses, equipment failure, operational unreliability and the incompliance of safety requirements could result from such problems. Thus solving these problems no matter the complexity becomes a priority. CI often requires minimal cost besides computational resources and technical skills of the engineer. The additional advantages of CI have caused many sectors (private and government) in the energy industry to adopt these measures (Fang and Lahdelma, 2016; Zhang *et al.*, 2016a; Mehdinejad *et al.*, 2017; Le Anh *et al.*, 2016).

Figure 1. Overview of various energy systems delivering electricity to the grid

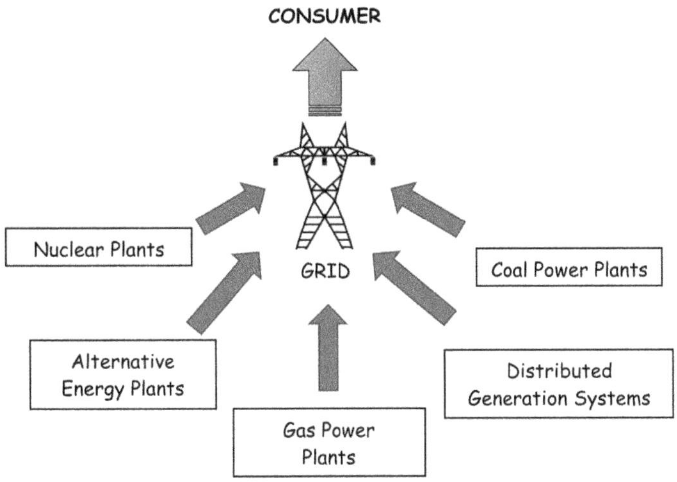

This chapter aims to provide a review on the current applications of CI in energy generation covering the main areas of energy systems. This chapter is organized as follows; in Section 2 of this chapter, a brief overview of recent works related to CI in engineering is presented. The subsequent subsections (Sections 3 to 8) comprise of the application of CI in various sectors of power systems including; renewable energy systems, distributed generation (DG), nuclear plants, coal power and gas-fueled plants. This chapter ends with some concluding remarks and recommendations for potential research directions on computational intelligence in energy generation.

COMPUTATIONAL INTELLIGENCE IN ENGINEERING

The broad field of CI could be divided into two categories based on the type of implementation often seen in engineering systems (see Figure 2).

The first category is predictive modeling; sometimes known as machine learning, these techniques are used to model highly complex problems (Kotsiantis *et al.*, 2007). Two main predictive modeling techniques frequently employed are artificial neural nets (ANN) and support vector machine (SVM) (also known as support vector regression (SVR)). The primary industries targeted by machine learning developers in the past year are shown in Figure 3.

It is shown in Figure 3, Utilities and Energy sector is among the top five industries targeted for machine learning.

Figure 2. Current CI techniques in engineering

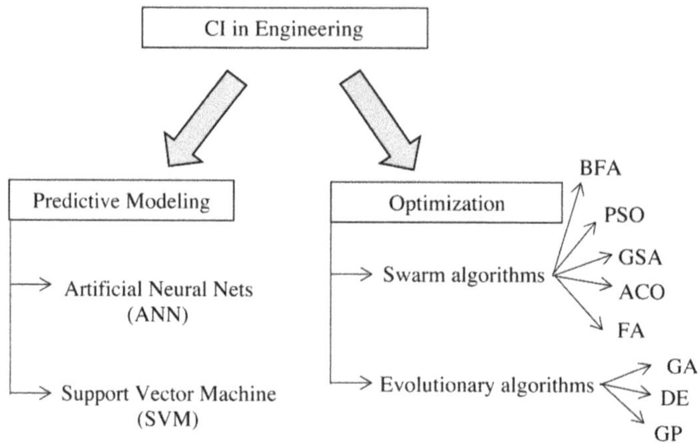

The second category in CI is techniques used for optimization - also known as metaheuristics. Metaheuristic techniques comprise of evolutionary and swarm-based algorithms as well as other nature-inspired approaches (Yan, 2010). Examples of swarm-based techniques are Particle Swarm Optimization (PSO) (Kulkarni *et al.*, 2015), Gravitational Search Algorithm (GSA) (Ganesan *et al.*, 2014), Ant Colony Optimization (ACO) (Mahi *et al.*,

Figure 3. Industries targeted by machine learning developers (Columbus, 2016)

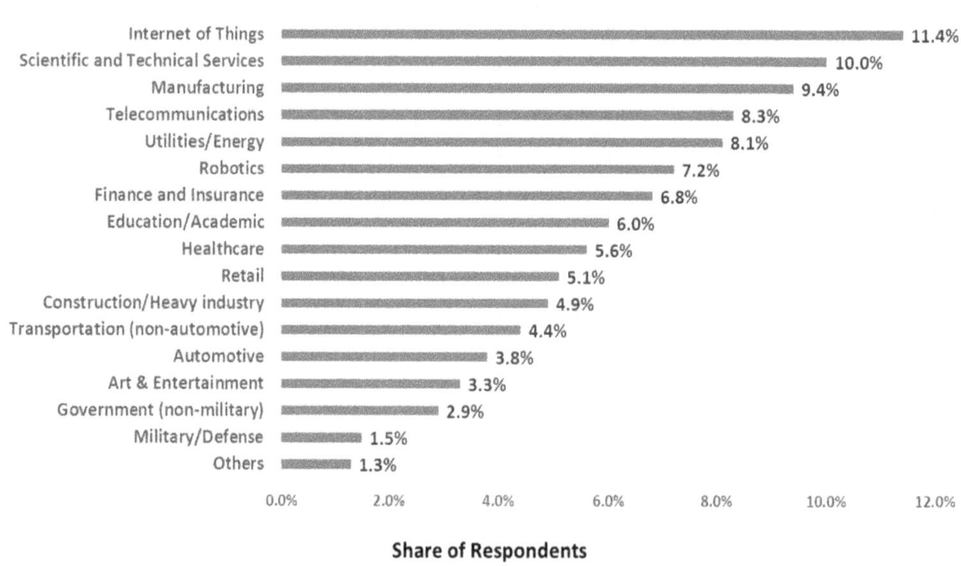

2015), Bacteria Foraging Algorithm (BFA) (Ganesan *et al.*, 2016) and Fish Swarm Optimization (FSO) (Soomro *et al.*, 2013). On the other hand, Genetic Algorithm (GA) (Deng *et al.*, 2015), Differential Evolution (DE) (Ganesan *et al.*, 2015) and Genetic Programming (GP) (Langdon and Harman, 2015) are among the most actively used evolutionary algorithms in engineering.

One popular and robust method for predictive modeling via CI is using ANN (Rafiq *et al.*, 2001; Chen *et al.*, 2017; Fayek *et al.*, 2016). Figure 4 shows a typical structure of a back-propagating ANN:

The ANN structure above could be presented in a simple mathematical form:

$$y_i = f(w_{ij}x_i) \text{ such that } i, j \in Z^+[1, N] \tag{1}$$

where the output vector y_i is a function of the product of the input vectors x_i and the weight matrix, w_{ij}. The positive integers *i* and *j* represent the dimension of the vectors which relates naturally to the dimension of the weight matrix - with a maximum *N* dimensions. During learning, the back-propagation mechanism (see Figure 1.4) updates the weights in the ANN for a fixed set of input and output vectors. An example of an update rule using stochastic gradient descent is given as follows:

$$w_{ij}(t+1) = w_{ij}(t) + l\frac{\partial C}{w_{ij}} + G(t) \tag{2}$$

Figure 4. Structure of a simple three-input, three-output back-propagating ANN with two hidden layers

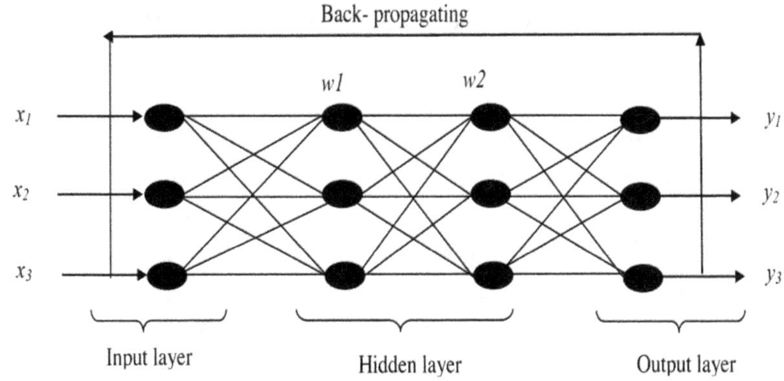

where the output vector l is the learning rate, C is the cost function and $G(t)$ is the stochastic term.

Over the past years neural nets have been employed extensively in various engineering industries. For instance, in material engineering, Esfe *et al.*, (2015) developed an ANN to predict the thermal conductivity and dynamic viscosity of ferromagnetic nanofluids. They used experimental data (temperature, diameter of particles, and solid volume fraction) as an input to the neural net (for training purposes). Using the network, the authors formulated two sets of correlations for their predictions.

Neural nets have also been extensively employed in structural engineering. In Chojaczyk *et al.*, (2015), the authors provided a comprehensive review on the usage of ANNs for performing structural reliability analysis. In that survey, the authors categorized various types of ANNs typically employed for problems involving structural reliability. They also elaborated on the types of ANN predictive models applied to structural design and optimization scenarios. In that work, the authors highlighted other methods that have been used in tandem with the ANN when solving such problems (e.g. Monte Carlo simulation (Gamerman and Lopes, 2006), first order reliability methods and Monte Carlo with importance sampling). Besides material and structural engineering, ANNs have also found applications in direct manufacturing. In Karabulut, (2015), ANNs and regression analyses was employed to predict the surface roughness and cutting force of an AA7039/Al_2O_3 metal matrix composite. The ANNs managed to predict these parameters with an accuracy of 2.25% (surface roughness) and 6.66% (cutting force). Another interesting research on surface roughness prediction was carried out by Khorasani *et al.*, (2015). In that work the authors predicted the surface roughness online during metal cutting operations using an ANN. The central idea of that work was to develop a dynamic monitoring system which simulates a wide range of machining conditions with and without the cutting fluid.

Applications of ANNs have recently extended to process engineering. Funes *et al.*, (2015) provides an in-depth review of ANNs used for modeling purposes of processes in the food industry. Besides they also broadly explained the main components of conventional ANNs. Another application of ANNs in the food industry is seen in Fan *et al.*, (2013). In that work, the authors used a back-propagating ANN for simulating and predicting the hardness and gumminess of food samples. They went on to prove that the ANN provided a better prediction in comparison with another simpler approach; linear fitting technique. An interesting application of ANNs in the food industry could also be seen in the works of Silveira *et al.*, (2013). In that work the authors predicted the final temperature of chicken carcass using an ANN on an industrial scale. They obtained the best prediction using an ANN with two hidden layers, radial bias transfer function and a gradient descent back-propagating algorithm (for training). Their results showed that the multivariable problem was effectively modeled using the ANN.

Jiang *et al.*, (2016a) employed a type of ANN (active deep neural net) for process engineering. They used the ANN to learn and perform fault diagnosis for chemical processes using chemical sensor data. The results from their research showed that the proposed approach obtained superior diagnostics accuracy with minimal data requirement. In fault diagnostics, ANNs have also been used on the Tennessee-Eastman process - which is a process control and monitoring method (Xie and Bai, 2015). In Xie and Bai, (2015), the authors used a Hierarchical Deep Neural Network which they trained to classify whole faults into a few groups. They compared the fault diagnosis results achieved using different types of neural networks.

Support vector machine (SVM) is an alternative method to ANN utilized for predictive modeling. However, SVM is more suited for data classification (a certain type of predictive modeling) as compared to ANN. Due to its effectiveness; this method has found many applications in engineering (Salcedo-Sanz *et al.*, 2014). Figure 5 presents the main components in the algorithm of the SVM approach.

SVM has been used with great success particularly in image processing and pattern recognition. For instance in Zhang *et al.*, (2015), SVM has been used for image classification of brain magnetic resonance images (MRI). In this type of implementation, the SVM (or ANN) is trained on a set of data where it learns how to classify specific types of data into their respective groups. When confronted with new data, the technique then performs the

Figure 5. Simplified algorithm of SVM

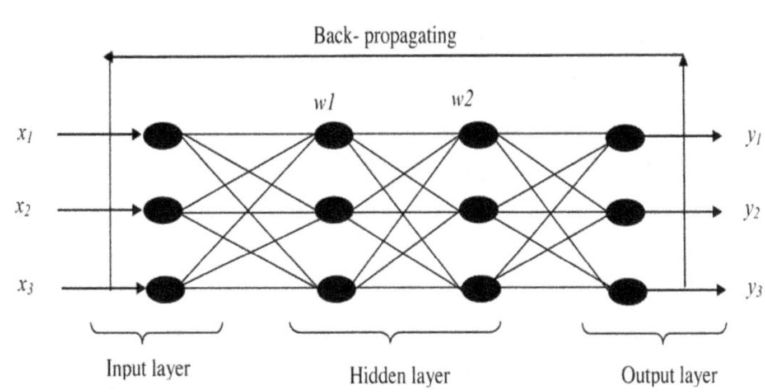

classification by predicting the appropriate group for this new set of data. This is done based on the predictive model developed during its training (Witten *et al.*, 2016). In the work by Zhang *et al.*, (2015), the authors combined SVM with other methods such as principal component analysis (PCA) and weighted-type fractional Fourier transform. In Bhat *et al.*, (2016), SVM was used in an image processing context for condition monitoring. By classifying the images of the tools, the SVM was used to determine the condition of the tool – how deeply it was affected by wear and tear from usage. The authors found that operating online, the SVM reliably classified tools making it cost-effective.

As with ANN, SVM has also been utilized in fault-detection applications. In Chang *et al.*, (2014), the authors used SVM to develop failure models of their cloud monitoring system. The cloud monitoring system was used for solar plants incorporating a geographic information system, an instantaneous power-consumption information system, a reporting system, and a failure diagnostics system. Another application in fault-detection was given in the work of de Souza *et al.*, (2014). Using SVM the authors studied fault scenarios on a cyclopentenol production reactor. The authors were interested in fault scenarios related to process efficiency and industrial safety. These scenarios were successfully investigated using SVM along with other methods such as PCA.

In the past few years, CI has also had a significant impact in engineering optimization. Besides predictive modeling approaches such ANN and SVM, optimization strategies using metaheuristics such as PSO, DE and GSA have become an active area of research. In robotics and microelectronics, swarm-based approaches such as PSO have been used as a tool to optimize design, operations and control (Singh *et al.*, 2015; Ranjani and Murugesan, 2015). This can be seen in Zhang *et al.*, (2014); where path planning for intelligent robots was optimized using PSO. PSO was combined with the grid method to enhance the performance of the traditional PSO. Their targets were to improve the real-time security and optimization ability of the path planning strategy by the intelligent robot. Another instance involving path planning optimization of robots could be seen in the work of Hossain and Ferdous, (2015). The authors in that work designed their BFA-inspired strategy using the C programing language in an OpenGL environment. To benchmark their approach, they compared the performance of their proposed approach against the PSO algorithm.

An interesting application of both categories of CI (predictive modeling and optimization) for robotics and control is given in the work of Qi and Chai (2014). In that work the PSO algorithm was combined with an ANN. Using a radial basis function ANN, the combined approach was used to model and compensate (optimize) for the error measurement of the 3D scanning robot. The authors stated that the proposed combined strategy effectively reduced the errors in model development and significantly increased the accuracy of the compensation. A similar application in control systems can be seen in the work of Precup *et al.*, (2013). In that work, the authors used an adaptive GSA for tuning their fuzzy controller attached to their servo systems. They tested their controller efficiency experimentally via position control of a laboratory servo system. Swarm-based approaches have also been used for tuning controllers in alternative energy systems. For instance in Chen *et al.*, (2015), a sun tracking control system used for increasing the output power of solar panels was optimized. Using the PSO approach in combination with the Taguchi Method and Logistic Map, the PI controller parameters of the tracking system was improved. The authors of Chen *et al.*, (2015) found that the improved tracking system enabled the solar panel to reach its maximum power point in a rapid and stable fashion.

In many engineering applications, cascade control has been observed to be difficult to tune due to its increased system complexity. Thus, implementing a metaheuristic for such cases may significantly help as shown in Saleem *et al.*, (2015). In that work the authors tuned the cascade control system which controls the servo-pneumatic system using PSO. The performance of this tuned cascade controller was tested experimentally under various conditions (speed variations). They showed that the PSO-tuned cascade control system performed significantly better as compared to the self-tuned controller. In addition to PSO, evolution-based techniques like the DE algorithm have also been extensively used in control engineering. An example of this sort of system could be seen in the work of Juang *et al.*, (2015). In that work the wall-following controls of a hexapod robot were optimized using a modified version of the DE algorithm called adaptive group-based DE. The DE approach optimized the fuzzy controller enhancing its efficiency and performance.

The DE algorithm was also applied to the intelligent flight control system of an unmanned aerial vehicle in Santoso *et al.*, (2017). The DE-improved control system was targeted to enhance its maneuvering and navigational capabilities via self-learning. In Senkerik *et al.*, (2014), the segments of the DE algorithm was modified for tuning controllers which are controlling systems exhibiting chaotic dynamics. The authors successfully applied their version of DE to optimize three different chaotic control systems. Besides regular control systems, the application of evolutionary algorithms has also been extended to the controls of quantum dynamics – which is a critical component in the field of quantum computing. An effort in this direction could be seen in the work of Zahedinejad *et al.*, (2014). Instead of relying on non-metaheuristic approaches such as greedy optimization (Naderi *et al.*, 2014), the authors of that work employed DE and PSO algorithms to deliver effective quantum control over their systems. When executed with the greedy approach, the DE and PSO techniques showed significant improvements in comparison.

It can be seen that recently CI has been widely implemented across various engineering disciplines – e.g. material engineering, structural engineering, manufacturing, process engineering, image processing, pattern recognition, robotics and control systems. For more comprehensive literature regarding recent advances in CI in engineering, see Yang *et al.*, (2016), Ganesan *et al.*, (2016), Senvar *et al.*, (2013) and Xiong *et al.*, (2015). In the following sections, further developments in CI specific to certain sectors of power generation are presented.

RENEWABLE ENERGY SYSTEMS

Solar Energy

As the negative effects of fossil fuels are felt around the world, many countries have designed power generation facilities which are fueled completely by alternative or renewable energy sources (e.g. solar, wind, biomass, hydroelectric or wave energy). To this end, CI-based technologies have been applied at various levels (Jha *et al.*, 2017; Vyas *et al.*, 2017).

In solar power plants, ANNs have been utilized to forecast solar irradiation (Watetakarn and Premrudeepreechacharn, 2015). Forecasting solar irradiation data is critical for the design and sizing of solar power facilities. Similarly the work of Verma *et al.*, (2016b) utilized an ANN to forecast the power generation from a solar plant - where they compared their approach with other forecasting techniques. These techniques include multiple linear regression, logarithmic regression and polynomial regression. Their model also accounted for parameters which could influence the degradation of their solar panels (e.g. cloud cover, temperature, wind speed, rainfall and humidity). Similar works on forecasting solar power/irradiation using ANNs could found in Hernández-Travieso *et al.*, (2014), Qazi *et al.*, (2015, Ferrari *et al.*, (2013) and Singh *et al.*, (2013). ANNs have also been used to simulate the power output of photovoltaic panels in solar plants. In Saberian *et al.*, (2014), the authors trained a general regression ANN and a feed-forward back-propagation ANN on 5-years' worth of data before using it to approximate the generated power output. Khademi *et al.*, (2016) performed power prediction and thermo-economic analysis on a solar power plant. They predicted their power output using a hybridized algorithm; ANN with Artificial Bees Colony (ABC) (a type of swarm metaheuristic). They tested their model on Tehran University's photovoltaic power plant and found that the model indicated that the plant was economically feasible on both cloudy and sunny weather conditions. Although SVMs have been used primarily for classification purposes, it has proved to be very successful in modeling applications similar to ANNs. This can be seen in the works of Kazem *et al.*, (2016), Zeng and Qiao (2013), Wan *et al.*, (2015) and Wang *et al.*, (2015). Those works focused on the prediction of output power of solar-based power systems.

Among many of the areas in solar energy, maximum power point tracking is one that has the most potential for optimization efforts. This makes it highly suitable for the application of CI. Maximum power point tracking are a group of strategies used to find the point where maximum extraction of energy occurs. An example of this sort of work is seen in Seyedmahmoudian *et al.*, (2015). In that work, the authors employed a hybrid evolutionary algorithm (DE and PSO) to detect the maximum power point under partial shading conditions. To further prove the proposed optimization technique, they verified their findings against experimental results. A similar work was carried out in Babu *et al.*, (2015) where a modified PSO technique was employed for maximum power point tracking. Their technique was able to identify the maximum power point of a PV system under the influence of an extremely dynamics environment. Their technique had zero steady state oscillations, fast dynamic response and was easy to implement. In Ahmed and Salam (2014), a type of swarm-based approach called the Cuckoo Search (CS) algorithm was applied for maximum power point tracking of a large and medium-sized PV system. Upon implementation under partial shading conditions, they found that the proposed approach showed outstanding capability, good transient performance and fast convergence. A summary of the CI literature in solar energy is given in Table 1.

Wind Power

Besides solar technologies, CI has also found applications in wind energy systems. Similar to irradiation forecasting for solar applications, ANNs have been used to predict wind energy (Chitsaz *et al.*, 2015; Ramasamy *et al.*, 2015). Since wind speed is the key parameter for wind power output prediction, Wang *et al.*, (2016) used a back-propagating ANN enhanced using GA (similar to Li *et al.*, 2014) for wind speed prediction. They validated their model using data from a wind farm located in Inner Mongolia. Based on sensitivity analysis, the authors of Wang *et al.*, (2016) showed that the parameters of the ANN could be significantly improved with the help of the GA. Another type of ANN, the wavelet neural net has also been effectively applied for wind speed predication. For instance in Doucoure *et al.*, (2016), wind speed prediction at different times was carried out using a combined approach (wavelet ANN and multi-resolution analysis). The multi-resolution analysis was seen to enhance the prediction done by the ANN in addition to reducing the computational effort of the algorithm. Osório *et al.*, (2015)

Table 1. Summary of the CI literature in solar power

References	Method & Application	Analysis
Watetakarn and Premrudeepreechacharn, (2015), Verma *et al.*, (2016b), Travieso *et al.*, (2014), Qazi *et al.*, (2015), Singh *et al.*, (2013).	ANN/Solar Irradiation Forecasting	ANNs show superior performance as compared to conventional techniques (e.g. multiple linear regression).
Ferrari *et al.*, (2013).	k-NN, ELM & SVR/Solar Radiation Prediction	SVR shows better performance as compared to others in terms of accuracy and computational cost.
Saberian *et al.*, (2014)	ANN/Solar Power Output Simulation	Extensive historical data learning prior to approximating the generated power output.
Khademi *et al.*, (2016)	Hybrid ANN-ABC /Thermo-economic modeling	Model indicated that the plant was economically feasible on both cloudy and sunny weather conditions.
Seyedmahmoudian *et al.*, (2015), Babu *et al.*, (2015), Ahmed and Salam (2014)	DE, CS and PSO/Power point tracking	Results showed zero steady state oscillations, fast response and easy to implement; good transient performance and fast convergence.
Kazem *et al.*, (2016), Zeng and Qiao (2013), Wan *et al.*, (2015) and Wang *et al.*, (2015)	SVM/Solar Power Output Simulation	Extending the application of SVM for modeling.

hybridized the ANN with several different approaches for wind data prediction. They combined mutual information, wavelet transform, evolutionary PSO, and the adaptive neuro-fuzzy inference system and compared their prediction with established methods. A current and comprehensive review on ANNs in wind energy systems is available in Alta (2015).

Along with ANNs, SVM techniques have also been implemented for wind speed forecasting. For instance in Santamaría-Bonfil *et al.*, (2016), short-term wind forecasting was done using a GA-enhanced SVM. The technique was implemented after wind speed data was mapped to a higher dimensional space via phase space reconstruction. The authors of that work performed a comparative analysis against other autoregressive approaches such as autoregression (AR), autoregressive moving average (ARMA) and autoregressive integrated moving average (ARIMA). The autoregressive models were tuned by Akaike's Information Criterion and the Ordinary Least Squares Method. The results of the comparison showed that their proposed approach performed a more accurate prediction of wind data. In Ren *et al.*, (2016), a novel technique combining SVM and empirical mode decomposition was proposed for wind speed forecasting. The authors in that work compared their predictions against real wind speed data and found that their approach outperformed several recent approaches in terms of accuracy. A different take on short-term wind speed prediction could be seen in the work of Chen and Yu, (2014). There the authors formulated the SVM model in a nonlinear state space. Using the unscented Kalman filter, dynamic state estimation was done on the wind sequence under stochastic uncertainty. They then went on to compare the prediction achieves by their approach with other autoregressive and ANN techniques.

Turning to the optimization aspect of wind energy, many state-of-the-art approaches have been applied to wind power systems. One key area of optimization is the layout of the wind turbine in the wind farm (Pookpunt and Ongsakul, 2013). In the work of Gao *et al.*, (2015), a multi-population GA was developed to obtain the most optimal wind turbine layout that maximizes the power output while minimizing the overall investment cost. In addition to comparing their approach with past works, they used their approach to assess the offshore wind power potential of Hong Kong. Another work on wind turbine layout optimization was done in Park and Law (2015). There the authors used a more classical and simple optimization approach for tackling their problem: sequential convex programming (SCP). Before application, they simplified their problem to a continuous and smooth function with convex level of nonlinearity so that the SCP technique could be applied successfully. Note that without such simplification, optimization is better performed using metaheuristics. For example in Hou *et al.*, (2015) as well as Feng and Shen (2015) wind layout optimization were done using PSO and the random search (RS) algorithms.

As with the wind turbine layout, optimization of the turbine itself is similarly critical when it comes to optimizing these power systems. This naturally becomes another avenue for the application of optimization techniques. An effort in this direction could be seen in the work of Mortazavi *et al.*, (2015). There a multiobjective GA was employed to optimize the geometrical characteristics of the airfoil sections of the wind turbine. They developed their models using computational fluid dynamics (CFD) and ANN. The optimized models were studies in terms of exergy analysis. A similar work was carried out by He and Agarwal, (2014) - where the authors optimized the airfoil of a wind turbine blade using a multiobjective GA. Using CFD as a supplement, the optimization was targeted to improve the turbine's lift to drag ratio.

Recently, optimization efforts have also been focused on the control engineering of wind turbine systems. This can be seen in the work of Petković *et al.*, (2014). In that work, an intelligent controller (using a neuro-fuzzy inference system) was designed and employed to ensure maximal power output of the wind turbine. The controller adjusts the system's speed such that it operates at its highest efficiency point. In Kahla *et al.*, (2015), the PSO algorithm was used to generate an on-off controller for a wind turbine system. The controller bases its operations on a maximum power point tracking mechanism. Controlling the rotor side converter of wind turbine, the optimized control system aims to maintain the wind turbine at its optimum power point. A proportional-integral (PI) controller was optimized in Sheikhan *et al.*, (2013). In that work the PSO coupled with a fuzzy logic technique was utilized to tune the PI controller for optimizing the wind turbine operations. Another PSO-based strategy was implemented by Hodzic and Tai (2016) to tune a wind turbine PI controller. Their PSO technique was combined with a Grey Predictor approach to enhance the controller performance. A similar strategy was employed to optimize the controls of a variable-speed, variable-pitch wind turbine in Jafarnejadsani *et al.*, (2013). Instead of using PSO or fuzzy-based approaches, the authors employed a radial-basis function ANN. They then validated their results using a 5 MW wind turbine simulator. Kusiak *et al.*, (2010) described a multiobjective optimization model of wind turbine performance. In that work, an evolutionary strategy algorithm is used to maximize the wind turbine power and minimize the turbine's drive train and tower vibrations. For a current and thorough review of performance optimization techniques implemented on wind turbines, see Chehouri *et al.*, (2015). A summary of the CI literature in wind energy is given in Table 2.

Bioenergy

CI has recently found applications in various aspects of bioenergy systems. ANNs has been a popular approach for predictive modeling in this field. This can be seen in the works of Farobie *et al.*, (2015) and Puig-Arnavat *et al.*, (2013). In Farobie *et al.*,(2015), an ANN was employed to predict biodiesel production of a supercritical non-catalytic spiral reactor. Comparing their predictions with experimental data they found that the ANN provided a very good estimate of biodiesel yield. The authors of Puig-Arnavat *et al.*, (2013) carried out a similar work by using an ANN to determine the gas yield and composition of a biomass gasifier. Using some input operating parameters of the gasifier as well as the biomass composition, the ANN successfully predicted the yield and composition of the fluidized bed reactors. A similar

Table 2. Summary of the CI literature in wind power

References	Method & Application	Analysis
Chitsaz *et al.*,(2015), Ramasamy *et al.*, (2015), Wang *et al.*, (2016), Li *et al.*, (2014), Doucoure *et al.*, (2016), Osório *et al.*, (2015), Alta (2015)	ANN/Wind Energy & Speed Prediction	Models were validated with real data and results were backed-up by sensitivity analysis. Combined approach (with wavelet and multiresolution analysis) is seen to be very effective.
Santamaría-Bonfil *et al.*,(2016), Ren *et al.*, (2016), Chen and Yu, (2014)	SVM/Wind Energy & Speed Prediction	Techniques used for data mapping to higher dimensional space. Combined techniques with unscented Kalman filter and empirical mode decomposition performed well.
Pookpunt and Ongsakul, (2013), Gao *et al.*, (2015), Park and Law (2015), Hou *et al.*, (2015), Feng and Shen (2015)	PSO, GA, SCP, RS/Plant Design Optimization – wind turbine layout	Successful multi -objective solution method. Problem simplification was carried out before solving with classical approach.
Mortazavi *et al.*, (2015), He and Agarwal, (2014)	GA, ANN/ Wind Turbine Geometrical Optimization	Works combined CI with CFD.
Petković *et al.*, (2014), Kahla *et al.*, (2015), Sheikhan *et al.*, (2013), Hodzic and Tai (2016), Jafarnejadsani *et al.*, (2013), Chehouri *et al.*, (2015)	ANN, PSO/Wind Turbine Control Optimization	CI effectively used for designing intelligent controllers and enhancing their performance.
Kusiak *et al.*, (2010)	ANN, ES/Wind Turbine PerformanceOptimization	CI used to model wind turbine performance and to optimize it.

work could be seen in Mikulandrić *et al.*, (2014) – where an ANN model was used to simulate fixed bed biomass gasifier reactors. The ANN models were further validated using empirically measured data. Hence ANNs have been recently seen to be very useful for modeling process kinetics (Casanova-Peláez *et al.*, 2015; Betiku and Taiwo, 2015; Maran and Priya, 2015; Betiku *et al.*, 2015).

The efficiency of biohydrogen production from sugar cane molasses has been modeled using ANN (Whiteman and Kana, 2014). In that work the authors utilized the ANN in tandem with the response surface methodology with the concentration of molasses, temperature and inoculum concentration as input parameters to the model. ANNs have also been used as a tool to model physical properties of various biodiesel samples. In Meng *et al.*, (2014), this kind of work was carried out where an ANN was used to predict the kinematic viscosity of biodiesel at the temperature of 313 K. A short review of current ANN applications in bioenergy research could be found in Sewsynker-Sukai *et al.*, (2017). In that work, recent findings on the implementation of ANNs for modeling and optimization of biohydrogen, biogas, biodiesel, microbial fuel cell technology and bioethanol systems are presented and discussed.

In addition to ANNs, SVM has also been implemented to bioenergy systems. In Shamshirband *et al.*, (2016), various versions of SVM approaches were applied to model a direct injection diesel engine. The diesel engine used diesel/biodiesel blends as fuel - containing expanded polystyrene wastes. They showed that some of the SVM variants were very successful in exergetic modeling of the combustion. Another work on biodiesel engine modeling using SVM is presented in Wong *et al.*, (2015). In that work a biodiesel engine was modeled and optimized using an ANN variant (extreme learning machine), CS as well as a least squares SVM. Besides engine modeling, SVM has also been a tool to determine (classify) the quantity of biodiesel constituent (Alves and Poppi, 2013). This is seen in Filgueiras *et al.*, (2014); where the quantity of animal fat biodiesel in soybean biodiesel and B20 diesel blends was determined using SVM and near infrared spectroscopy. Similar to ANNs, SVM has also been used to predict process yields. In Cao *et al.*, (2016), the biochar yield from cattle manure pyrolysis was predicted using ANN and least squares SVM. They validated their predictions using 33 experimental data sets obtained using a laboratory-scale fixed bed reactor. Their analyses showed that the least squares SVM performed better predictions compared to the ANN.

Metaheuristic strategies have also been used for optimization in many areas related to bioenergy. A comprehensive review on biomass-for-bioenergy supply chain optimization methods is given in De Meyer *et al.*, (2014), Castillo-Villar, (2014) and Yue *et al.*, (2014). In the work of Mirkouei *et al.*, (2016), the GA was employed to find solutions to a biomass-to-bioenergy supply chain model. The authors developed this stochastic multi-criteria model containing multiple uncertainties in its variables. Besides supply chain, optimization has also been done on the combustion aspects of biodiesel engines. For instance in Shirneshan *et al.*, (2016), a swarm approach; ABC algorithm was utilized to optimize engine performance and emission characteristics of biodiesel engine. This was performed by finding for the optimal percentage of fuel mixture (biodiesel blended with diesel), engine speed and engine load. They found that the ABC algorithm successfully found the optimal operational settings for the engine. Another work involving biodiesel engine optimization could be seen in Zhang *et al.*, (2016b). In that work, a comparative analysis between two techniques when applied to optimize a biodiesel engine was given. The authors compared the performance of a regular GA against that of the hybrid PSO-GA approach. They found that the hybrid algorithm performed superior than the regular GA strategy by further optimizing the dynamometer time, fuel economy, and exhaust emissions.

Besides engine optimization, an interesting avenue for optimization of bioenergy systems is bioresource management. In Herman *et al.*, (2016), the bioenergy crop selection and placement was optimized using GA such that it reduces the negative impacts on nearby water sources. A recent work on bioenergy production is done in Sebayang *et al.*, (2017). In that work the authors focused on optimizing the bioethanol production from sorghum grains. They modeled the process responses using an ANN which they then optimized with an ACO algorithm. The results of the optimization were reduced production cost, time and effort as compared to relying on solely on experimental techniques. The following works provides a comprehensive review on recent optimization works in bioenergy systems: Asadi and Sadjadi, (2017), Ba *et al.*, (2017) and Mirkouei *et al.*, (2017). A summary of the CI literature in bioenergy is given in Table 3.

Hydro and Hydro-Thermal Power

Among key applications of predictive modeling in hydropower is the implementation of ANNs in its control systems. This is seen in the work of Goyal *et al.*, (2014). In their work, an ANN was used to learn the best control parameters for controlling the speed of hydroturbines – flow control. The ANN

Table 3. Summary of the CI literature in bioenergy

References	Method & Application	Analysis
Farobie *et al.*, (2015), Puig-Arnavat *et al.*, (2013), Mikulandrić *et al.*, (2014)	ANN/Reactor Design Optimization	Studies showed that the model developed using ANN for reaction engineering agreed very well with experimental data.
Casanova-Peláez *et al.*, (2015), Betiku and Taiwo,(2015), Maran and Priya, (2015), Betiku *et al.*, (2015), Whiteman and Kana, (2014), Cao *et al.*, (2016)	ANN, SVM/ Process Modeling	ANN model was employed to effectively model the process kinetics.
Alves and Poppi, (2013), Filgueiras *et al.*, (2014),	SVM/Fuel blends Optimization	SVM performed classification and modeling – studies showed that in these applicaitons SVM performed better than ANN.
Meng *et al.*, (2014), Wong *et al.*, (2015), Shamshirband *et al.*, (2016), Shirneshan *et al.*, (2016), Zhang *et al.*, (2016b)	ANN, CS, ABC, PSO, GA/ Biodiesel Engine Performance Optimization	SVM variants were very successful in exergetic modeling of the combustion.
De Meyer *et al.*, (2014), Castillo-Villar, (2014), Yue *et al.*, (2014), Mirkouei *et al.*, (2016), Herman *et al.*, (2016), Sebayang *et al.*, (2017)	GA, ACO/ Bioresource Management Optimization	Hybrid algorithms performed better than stand-alone techniques; especially in multicriteria settings.

then chooses the optimal gains for tuning the controller at various conditions significantly improving the control response. Such control applications are mainly seen in combined hydro-thermal plants. An example is given in the work of Prakash and Sinha, (2015). There an ANN-fuzzy approach was used for generation control of a combined hydro and thermal power plant. The controller performance was simulated using the Matlab/Simulink software. The authors in Prakash and Sinha, (2015) also conducted a comparison study among three control strategies; ANN-based, fuzzy-based and the hybrid ANN-fuzzy approach.

Controllers in hydro-thermal plants could also be directly improved via the application of optimization techniques. In Kaliannan *et al.*, (2015), a metaheuristic technique was applied to enhance a load frequency controller. In that article, the authors presented the results from the implementation of the Ant Colony Optimization (ACO) algorithm. They used ACO to tune a proportional-integral-derivative (PID) controller used in a hydro-thermal plant. They compared their intelligent control strategy against the conventional PI controller – which was tuned via trial-and-error. The comparative analyses confirmed the superiority of their proposed method.

Inspired by Kaliannan *et al.*, (2015), Shivaie *et al.*, (2015) extended the study by using harmony search algorithm (HSA), a type of swarm technique for optimizing the load frequency control for a hydrothermal plant. Their work also tackled a more general problem; non-linear interconnected hydrothermal power system. Similar works on the application of metaheuristics for load frequency control optimization in hydro-thermal plants could be seen in Pappachen and Fathima, (2016), Prakash and Sinha (2014), Bhateshvar *et al.*, (2015). An intelligent frequency controller for the synchronous generator of hydro turbine is also described by Syan *et al.*, (2015).

In some engineering applications, improvement efforts require both classes of CI; predictive modeling and optimization. The predictive modeling part develops a working model from past/historical data. The optimization technique optimizes the developed model by finding the suitable parameters. Due to such robustness and strength, these hybrid approaches (predictive modeling – optimization) have become increasingly popular in many engineering as well as power system applications.

Such a scenario could be seen in Aboutalebi *et al.*, (2015) where the monthly operation rules for a hydropower reservoir operation is optimized. The optimization was carried out using a hybrid NSGA-II and SVR. They applied their approach to a real-world hydropower system; Karoon-4 reservoir of Iran. Upon application, the authors showed that they successfully optimized their reservoir with an accuracy of about 90%. Another hybrid implementation is presented in the work of Li *et al.*, (2013). In that work the GSA algorithm was modified using a kernel clustering method (basis of the SVM approach) for fault diagnosing a hydroelectric generating unit. They applied their strategy to a real system and compared their approach with traditional non-hybrid methods. The results confirmed the superiority of the proposed hybrid technique.

Besides, hybrid frameworks, approaches have also been developed purely for predictive modeling of hydropower systems. This is shown in Cheng *et al.*, (2015). There the authors used two ANN variants to perform daily runoff forecasting for a hydropower station. They assessed their method using real daily runoff data of the Hongjiadu reservoir in southeast Guizhou province of China from 2006 to 2014. Their tests showed that the ANN variant performed very well in predicting the reservoir runoffs. A similar effort towards modeling of hydropower systems is presented in Hammid *et al.*, (2017). In that work, an ANN was employed to predict the power production of a small-scale hydropower plant located in Himreen Lake, Iraq. The model developed using the ANN accounts for the uncertainties in operational input and output variables. The ANN was trained using 3570 points of experimental data obtained from the nonlinear plant. Their analyses showed that the model closely produced the power output of the small-scale hydropower plant. In addition to power output, reservoir water inflow is another critical variable which comes into consideration when designing and operating hydropower plants. This sort of predictive effort was done in Sauhats *et al.*, (2016). There the authors used an ANN to model critical reservoir parameters: temperature, precipitation and historical water inflow. Their study concentrated on a hydropower plant in Latvia.

Recently there have also been studies that directly apply optimization approaches to improve hydropower systems. In Moreno and Kaviski, (2015), the problem of optimal scheduling of power dispatch was considered. The authors approached the issue by implementing the PSO technique to generate feasible solutions which maximize the cascade of electricity production while accounting for the environmental and water balance constraints. The results of extensive computational experiments showed that their optimization strategy outperformed other heuristic techniques. A similar work on hydro-thermal power scheduling could be seen in Glotić and Zamuda, (2015) – where the the DE algorithm was employed to obtain optimal thermal schedules. The work aimed to find such schedules where the short-term economic and emissions are reduced. Four case studies were used to verify their proposed optimization procedures.

Besides power dispatch and scheduling, optimization of hydropower operations is also crucial. In Kiruthiga and Amudha, (2016), the PSO algorithm was used to derive optimal operational rules which maximize the output power of a hydroplant. The optimization was targeted towards the Aliyar reservoir in the Coimbatore district of Tamil Nadu, India. They obtained various reservoir release patterns which they then optimized using PSO - to maximize the generation capability of the plant. Similar works on the optimization of hydropower plant operations could be seen in Azizipour *et al.*, (2016), Zhang *et al.*, (2016c), Ghimire and Reddy, (2013), Yang *et al.*, (2015a) and Tayebiyan *et al.*, (2016). In those works the authors implemented variants of PSO, GA as well as a type of evolutionary technique, invasive weed optimization (IWO) algorithm. A summary of the CI literature in hydro- and hydrothermal energy is given in Table 4.

An interesting review of optimization and machine learning methods in renewable energy systems could be seen in Iqbal *et al.*, (2014). Figure 6 presents a general overview of current applications involving CI in alternative energy systems as previously discussed in this section.

Table 4. Summary of the CI literature in hydro- and hydrothermal power

References	Method & Application	Analysis
Goyal *et al.*, (2014), Prakash and Sinha, (2015), Kaliannan *et al.*, (2015), Kaliannan *et al.*, (2015), Shivaie *et al.*, (2015), Pappachen and Fathima, (2016), Prakash and Sinha (2014), Bhateshvar *et al.*, (2015)	ANN, ACO, HSA/Plant Controls Optimization	Intelligent controllers shown to have superior performance compared to conventional controllers. Metaheuristics was used to automate the tuning of the controllers.
Aboutalebi *et al.*, (2015), Li *et al.*, (2013)	NSGA-II, SVR, GSA/ Reservoir Operations Optimization	Reservoir operations were optimized with 90% accuracy. Hybrid techniques were successful in fault-diagnosing.
Cheng *et al.*, (2015)	ANN/Daily Runoff Forecasting	An ANN variant performed very well in predicting the reservoir runoffs.
Syan *et al.*, (2015)	ANN/Hydro Turbine Frequency Control	An ANN shows better control as compared to conventional controller
Hammid *et al.*, (2017)	ANN/Plant Power Output Prediction	ANN successfully predicted the power production of the plant. The ANN accounts for the uncertainties in operational input and output variables which contained uncertainties.
Sauhats *et al.*, (2016)	ANN/Reservoir Water Inflow Prediction	ANN modelled critical reservoir parameters: temperature, precipitation and historical water inflow.
Moreno and Kaviski, (2015), Glotić and Zamuda, (2015)	PSO, DE/Power Dispatch/Scheduling Optimization	The results of extensive computational experiments showed that their optimization strategy outperformed other heuristic techniques.
Kiruthiga and Amudha, (2016), Azizipour *et al.*, (2016), Zhang *et al.*, (2016c), Ghimire and Reddy, (2013), Yang *et al.*, (2015a) Tayebiyan *et al.*, (2016)	PSO, GA, IWO/ Hydropower Plant Optimization	Various reservoir release patterns were optimized using PSO to maximize the generation capability of the plant.

DISTRIBUTED GENERATION

In distributed generation (DG) systems, power is stored and generated by multiple different sources. This is in contrast with conventional power stations where energy is usually produced using a single fuel source. DG systems usually have multiple renewable energy sources coupled with a conventional source (diesel or natural gas) (Singh *et al.*, 2016; Huda and Živanović, 2017). The idea is to reduce the usage and dependency on fossil fuel by integrating multiple renewable energy systems into a single system. A general schematic of a DG system with multiple alternative energy sources is given in Figure 7.

Nevertheless DG systems still have a few setbacks which engineers are trying to resolve – e.g. cost effectiveness, reliability and emissions (Ganesan

Figure 6. Overview of current applications involving CI in alternative energy systems

- Thermo-economic modeling
- Power point tracking

WIND POWER

- Wind Energy/Speed Prediction
- Plant Design Optimization – wind turbine layout
- Wind Turbine Geometrical Optimization
- Wind Turbine Control Optimization

BIOENERGY

- Reactor Design Optimization
- Process Modeling
- Fuel blends Optimization
- Biodiesel Engine Performance Optimization
- Bioresource Management Optimization

HYDROPOWER

- Plant Controls Optimization
- Reservoir Operations Optimization
- Daily Runoff Forecasting
- Plant Power Output Prediction
- Reservoir Water Inflow Prediction
- Power Dispatch/Scheduling Optimization
- Hydropower Plant Optimization

Figure 7. Simple schematic of a DG system

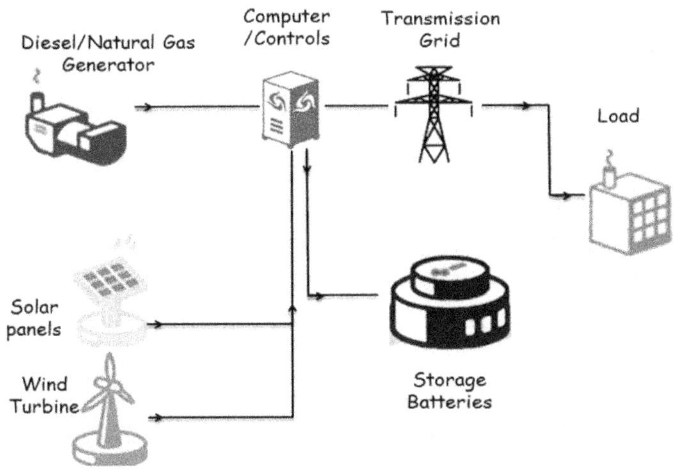

et al., 2013; Georgilakis and Hatziargyriou, 2013; Wen *et al.*, 2015). A typical depiction of the objective function for design cost of a DG system is as follows:

$$COST = \frac{\sum_{i=w,s,b}(I_i - S + OM_i)}{N_p} + C_g \quad (3)$$

where COST is the design cost in \$/year while the subscripts *w*, *s*, *b* stand for wind, solar and battery storage respectively. I_i represents the initial cost, *S* is the salvage values OM_i is the operations and maintenance costs. C_g is the annual cost of purchasing from the utility grid.

These setbacks are mostly due its system complexity and thus involve controls, design and operations. The book by Mahmoud and Fouad, (2015) highlights these issues and presents how microgrids, interconnected groups of energy units and localized loads could balance power supply and demand effectively. The book also explores various aspects of DG systems including system architecture and control. System integration, modeling and analysis as well as communications are also discussed in a practical sense.

A recent work applying CI to DG systems could be seen in Elkazaz *et al.*, (2016). In that work, the authors focused on optimizing the operations of a DG system – which comprises of a hybrid fuel cell and photovoltaic (PV) system for producing power to a residential area. Using genetic algorithm (GA) (an evolution-based metaheuristic), the authors optimized the DG system significantly improving its operations and consequently the operating costs. In Kumar and Thakur, (2014), a comparative analysis of a series of PSO variant algorithms applied to DG systems was carried out. The authors used the techniques to identify the most optimal location and sizing of a single DG system such that it minimizes the total active power loss in the system. In the same spirit the authors of Rao *et al.*, (2013) applied CI with the target to minimize power loss in their DG system. Unlike the work of Kumar and Thakur, (2014), the optimization in that work was done to improve the voltage profile distribution in the system using HSA. Inspired by the improvisation of jazz musicians, HSA has been proven to be a highly effective metaheuristic in many classes of optimization problems (Manjarres *et al.*, 2013). Similar optimal DG allocation and sizing using metaheuristics could be seen in the works of García and Mena, (2013), Kansal *et al.*, (2013), Abdi and Afshar, (2013), Moradi *et al.*, (2014) and Kowsalya (2014).

Besides allocation and sizing, another crucial aspect of DG systems is contract pricing. In this setting, the two parties involved are the owner of the DG units and the distribution company. The distribution company aims to maximize profits by minimizing payments made to the owner. In return, the owner of the DG units aims to increase its energy sales (Sadeghi *et al.*, 2016). This scenario is reflected in the work of Hejazi *et al.*, (2013); where the authors propose a type DE approach which determines the best locations, sizing and payment incentives for the DG system. Their objective of the algorithm is to maximize the profit of the distribution company by procuring power at an optimal price to meet the market demand. At the same time the algorithm ensures that the investment remains attractive to owner of the DG units. The proposed DE approach strikes an economical balance between both parties to achieve an optimal contract pricing. A similar work is seen in Ameli *et al.*, (2015). The authors used the multiobjective PSO to simulate various scenarios on an IEEE 33-bus distribution test system. From the simulated results, they selected the solution for the DG system's location, sizing and contract price. The proposed technique accounts for power loss reduction, voltage profile and stability improvement and reliability enhancement as well as the previously mentioned economical elements. Thus the delicate element of DG optimization lies in the balancing of the engineering (location and sizing) against the economic incentives between the parties involved. The key implementation areas of CI in DG systems are:

- Plant location selection
- Plant and component sizing

- Power output and reliability improvement
- Optimal Contract pricing
- Engineering-economic balancing

A summary of the CI literature in DG systems is given in Table 5.

NUCLEAR PLANTS

Nuclear technology in the power industry has significantly gained from the developments in predictive modeling – particularly in the development of the SVM technique. This can be seen in the field of predicting power consumption by nuclear power plants; explored in the work by Kleshchyova *et al.*, (2015). In that work the authors discuss the financial and operational importance of predicting the nuclear plant power consumption. Due to those reasons, the authors employed SVM for high-accuracy prediction. SVR was also used in Liu and Zio (2015) for system diagnostics under non-stationary conditions. The work focused on applications in predicting leakages in the first seal of a reactor coolant pump of a nuclear power plant.

SVR was also employed in Liu *et al.*, (2013a) to predict nuclear power plant failure scenarios. The inputs to the model were selected using Fuzzy Similarity Analysis (FSA) and the prediction was carried out on a drifting process parameter of a nuclear power plant component. Another fault diagnostics application could be seen in Yu *et al.*, (2016). The authors in that work tailored their application towards the feedwater pump in the secondary circuit - which is critical to the operations of the nuclear plant. Comparative

Table 5. Summary of the CI literature in DG systems

References	Method & Application	Analysis
Kumar and Thakur, (2014), Ganesan *et al.*, (2013), Elkazaz *et al.*, (2016), Rao *et al.*, (2013)	ANN, PSO, GA, HSA/DG Plant design, allocation and component sizing	Optimal location and sizing of a single DG system using a variety of CI techniques.
Sadeghi *et al.*, (2016)	DE/Optimal Contract pricing	Metaheuristic effectively used to maximize profits by minimizing payments made to the owner of the DG system.
Hejazi *et al.*, (2013), Ameli *et al.*, (2015)	DE,PSO/Engineering-economic balancing	The techniques accounts for power loss reduction, voltage profile, stability improvement, reliability enhancement as well as the economical elements.

analyses was carried out against another method; the probabilistic neural network. They showed that the SVM produced results that were more accurate and effective in diagnosing the faults of the feedwater pump. In Liu *et al.*, (2013b) the implementation of SVR was extended to component condition monitoring in nuclear plants. A modified version of a SVR was employed and its parameters were tuned. Model identification, uncertainty analysis and comparison studies were also carried out in Liu *et al.*, (2013b). A similar implementation of SVR on condition monitoring was presented in Li *et al.*, (2017). Using kernel entropy component analysis and SVM, the authors designed a fault diagnostic framework for a wireless sensor network in a nuclear plant. The wireless sensor network is used for condition monitoring of the components of the plant. A more extensive and rigorous work on condition monitoring in nuclear power plants can be seen in Jiang *et al.*, (2016b). The authors of that work looked at condition monitoring via accurate prediction of nuclear plant behavior during severe accident scenarios - after initiating events such as rod ejection accidents (REA) and rod drop accidents (RDA). They performed the study using SVR as well as an improved SVR called the fuzzy-weighted SVR. Comparing their approaches with the back propagating ANN, the authors proved the superiority of the fuzzy-weighted SVR.

ANNs have also been employed in nuclear plants in the area of control engineering. In de Oliveira and de Almeida, (2013), an ANN was used to model the pressurized water reactor. The pressure control of the mentioned reactor is critical for maintaining the safety condition and the generation efficiency of the nuclear plant. Once a model was developed using ANN, a fuzzy controller was employed to ensure the set point of the pressurized water reactor. In their simulations, they showed that their ANN model worked accordingly and the fuzzy controllers performed much better as compared to conventional PID controllers. A similar work involving the controls of a nuclear reactor could be found in Dong (2014). There the main focus was on the control system of the high temperature gas-cooled reactor. The author of that work used the ANN to compensate for uncertainties in the proportional-derivative (PD) power level controller. Due to the complexity of the pressurized water reactor control system, research efforts have been directed towards the tuning of the control system (as done in Vadivel *et al.*, 2014). In that work, the controller of the reactor was tuned using two optimization techniques; GA and PSO. The authors of that work found that the PID controller tuned this way was more effective than the ones tuned using the conventional Ziegler-Nichols method. In Coban, (2014) a multi-layered feedback ANN was employed to design a new controller for neutronic power level control of a nuclear research reactor. The ANN was trained with the help of the PSO algorithm. Learning at different power levels, the proposed controller was seen in simulations to have excellent performance especially for tracking step reference power trajectories. An in-depth study of application of ANNs in nuclear engineering could be seen in Cong *et al.*, (2013). Besides ANN, control systems in nuclear plants have also been optimized by direct application of metaheuristic techniques (Santhiya and Pappa, 2014).

In addition to predictive modeling, optimization techniques have also been applied to nuclear power plants. Metaheuristics techniques have been recently applied to nuclear safety systems. In Di Maio *et al.*, (2014) the Hierarchical DE for identification of minimal cut sets involving coherent and non-coherent fault trees. The Hierarchical DE was applied to two systems:

1. The reactor protection system of a pressurized water reactor.
2. The airlock system of a Canadian Deuterium Uranium (CANDU) reactor.

The authors in Di Maio *et al.*, (2014) analyzed their solutions in terms of accuracy and computational demand. Optimization has also been done on the design of core fuel loading patterns in nuclear reactors. For instance in Karahroudi *et al.*, (2013), the optimal design of fuel core loading patterns was obtained using GA. Analyzing the results for a VVER-1000 reactor, the authors found that the responses of the obtained optimal loading pattern was acceptable and within the defined boundary conditions. Further works on optimization of core fuel loading using swarm-based algorithms and other metaheuristics are seen in Augusto *et al.*, (2015), Poursalehi *et al.*, (2013a), Poursalehi *et al.*, (2013b), Poursalehi *et al.*, (2013c), Poursalehi (2015), Jayalal *et al.*, (2014), Jamalipour *et al.*, (2013), Hill and Parks (2015), Mortezazadeh *et al.*, (2015), Schlünz *et al.*, (2015), Mahmoudi *et al.*, (2016) and Kashi *et al.*, (2014).

Besides fuel core loading, reactor design optimization has also been done using metaheuristics. An example of this could be seen in Sacco and Henderson (2014). There a modified DE algorithm was introduced to solve the nuclear reactor core design optimization problem. The design optimization problem involved several reactor cell parameters such as: cell dimensions, enrichment levels and materials. The work was targeted to minimize the average peak-factor in three-enrichment-zone reactor while accounting for constraints on average thermal flux, criticality and sub-moderation. The authors compared their proposed technique against the conventional DE approach - demonstrating the superior results of the proposed technique. Similar works on design optimization of nuclear reactor core using DE could be seen in Sacco *et al.*, (2013), Sacco *et al.*, (2014) and Sacco and Henderson (2015). Another design problem in a nuclear power plant was solved in Ibrahim *et al.*, (2014). There the PSO technique was used to obtain the optimal design of a U-tube steam generator. Since the problem was multiobjective in nature, the authors carried out optimization using another variant of PSO suited for this sort of problems; the non-dominated sorting PSO.

GAs have also been used for surveillance testing in nuclear plants (Mehra *et al.*, 2013). Surveillance tests are routinely performed on the plant's standby systems to ensure availability on demand – especially systems which are critical to safety. In Mehra *et al.*, (2013), the GA was applied to the problem to schedule these surveillance activities in a cost-effective way. In that work, various segments of the GA technique were explored and modified to ensure optimal performance. The current applications of CI in nuclear power plants are as follows:

- Plant Power Consumption Prediction
- Plant Failure Scenarios & Fault Diagnostics
- Condition Monitoring
- Control Engineering
- Nuclear Safety Systems
- Fuel Core Loading
- Nuclear Reactor Design
- Plan Surveillance Testing

A summary of the CI literature in nuclear plants is given in Table 6.

COAL POWER

As the global supply of high-quality coal deteriorates, various measures have been taken by energy providers to ensure consistent supply of power while maintaining a decent profit margin. Among these technologies are: implementation of coal blending, using higher efficiency boilers (supercritical boilers) and coal additives (Shui-jun *et al.*, 2012; Sloss, 2014). To help rapidly realize these technologies, these industries turn to CI. For example in Lv *et al.*, (2013), a novel variant of an SVM was employed to predict the NO_x emissions of a coal-fired boiler. The model was developed based on real-time operations data. The proposed SVM variant was enhanced with a

Table 6. Summary of the CI literature in nuclear plants

References	Method & Application	Analysis
Kleshchyova *et al.*, (2015)	SVM/Plant Power Consumption Prediction	Effectively employed SVM for high-accuracy operational and financial prediction.
Liu and Zio (2015), Liu *et al.*, (2013a), Yu *et al.*, (2016)	SVR/Plant Failure Scenarios & Diagnostics	SVR was implemented for system diagnostics under non-stationary conditions.
Liu *et al.*, (2013b), Li *et al.*, (2017), Jiang *et al.*, (2016b)	SVR, ANN/ Condition Monitoring	The techniques were coupled with fuzzy-based approaches to model uncertainties.
Oliveira and de Almeida, (2013), Dong (2014), Vadivel *et al.*, (2014), Coban, (2014), Cong *et al.*, (2013)	ANN, GA, PSO/Control Engineering	Techniques were used to design fuzzy controllers and to intelligently tune controllers.
Di Maio *et al.*, (2014)	DE/Nuclear Safety Systems	Fault tree identification was achieved where the solutions were analyzed in terms of accuracy and computational demand.
Karahroudi *et al.*, (2013), Augusto *et al.*, (2015), Poursalehi *et al.*, (2013a), Poursalehi *et al.*, (2013b), Poursalehi *et al.*, (2013c), Poursalehi (2015), Jayalal *et al.*, (2014), Jamalipour *et al.*, (2013), Hill and Parks (2015), Mortezazadeh *et al.*, (2015), Schlünz *et al.*, (2015), Mahmoudi *et al.*, (2016), Kashi *et al.*, (2014)	GA and swarm-based techniques/ Fuel Core Loading	A variety of metaheuristics were employed to continuously to obtain an optimal core loading pattern.
Sacco and Henderson (2014), Sacco *et al.*, (2013), Sacco *et al.*, (2014), Sacco and Henderson (2015), Ibrahim *et al.*, (2014)	DE, PSO/Nuclear Reactor Design	Demonstrated superiority of proposed technique used to minimize the average peak-factor in three-enrichment-zone reactor.
Mehra *et al.*, (2013)	GA/Plan Surveillance Testing	Various segments of GA was modified to ensure optimal performance.

combination of the least squares method as well as fuzzy membership grades. On the other hand, the authors of Zhang *et al.*, (2014) developed a hybrid approach for ash content prediction of coarse coal via image analysis. Ash content is one of the major indices used to measure coal quality. The authors performed feature extraction using GA and then applied the SVM to model the ash content. Using metrics like root mean squared error (RMSE) and

R-squared for accurate prediction of ash content, they established that the predictions on larger size fractions were better than smaller ones.

Certain works in coal plants do not only need modeling but also require optimization. In such situations a combined approach (consisting of both modeling and optimization) is often employed. This can be seen in the paper by Tan *et al.*, (2014) – where the authors worked on modeling and optimizing the NO_x emissions from a 700 MW pulverized coal-fired power plant. Using an extreme learning machine (Zong *et al.*, 2013) for modeling and HSA for optimization, their combined technique was aimed at optimizing the boiler's operating parameters. The optimization was targeted towards the minimization of NO_x emissions based on the model developed by the extreme learning machine. Their results showed that the proposed method was acceptable for online optimization of the boiler and capable of obtaining results within a single second. Along that same line of research, Liu *et al.*, (2015) optimized a coal-fired boiler using a different approach compared to Zong *et al.*, (2013). In Liu *et al.*, (2015), the authors used a conventional neural net to model the combustion data and a GA for optimization. They focused the optimizing their production by adjusting the air and fuel flow rates. Optimization has also been done on other parts of the coal power plant besides the combustion process. This can be seen in Yang *et al.*, (2015b). In that work, the authors focused on their study on coal slurry transportation pipelines. In such pipelines, blockage is a commonly occurring problem. To determine pipeline blockages, pressure prediction at the measuring points are very useful. The authors in Yang *et al.*, (2015b) used three hybrid methods for developing a predictive model:

- Extreme learning machine optimized with PSO (PSOKELM)
- SVM optimized with PSO (PSOSVM)
- Extreme learning machine optimized with the kernel function from SVM (KELM)

They found that the PSOKELM outperformed the PSOSVM and KELM techniques when compared using the mean squared error (MSE) metric.

In coal power plants, ash fusion temperature is another key indicator used for measuring the potential for a particular type of coal to slag. Slagging in coal boilers could pose a serious problem to the operations of the plant during power generation. Thus predicting the ash fusion temperature remains critical to this industry. In Karimi *et al.*, (2014) a type of neural network: adaptive neurofuzzy inference system was employed to predict ash fusion temperatures for US coals. The adaptive neurofuzzy inference system was trained on 1040 US coal samples from 12 states. The parameters considered in their work were initial deformation temperature, softening temperature and fluid temperature.

Besides ash fusion temperature, carbon content is another important measure used for optimizing and designing coal-fired boilers. In Dindarloo and Hower, (2015), the prediction of ash content was carried out using three approaches: multiple regression (MR), decision trees (DT) and SVM-based regression. Their work targeted to predict unburned carbon content in fly ash in a 300 megawatt (MW) power plant with a tangentially fired furnace. Using boiler operational data and coal petrological variables from past experiments, all three approaches managed to predict carbon content with reasonable accuracy although the SVM-based regression proved far more superior in terms of accuracy as compared to the MR and DT techniques.

Works purely involving optimization has also been done on coal power plants. In Wang *et al.*, (2014a), the design of an entire coal plant was optimized in a multiobjective setting using an enhanced DE. The authors focused on finding the optimal trade-off between thermodynamics and economics of the plant – since investment cost increases with thermal performance. In coal industries, supercritical boilers are more efficient in driving steam turbines as compared to regular boilers. Therefore further optimizing these boilers have been recently of some interest. The work by Wang *et al.*, (2014b) involved parametric optimization on such boilers by combining two techniques. The first was a classical optimization technique called mixed-integer nonlinear programming while the second was an improved DE algorithm. Similar to Wang *et al.*, (2014a), the authors performed an optimization to find the best trade-off between thermodynamics and economics. Implementing their combined technique, they found that it managed to decrease the cost of electricity by 2% for a single reheating unit. Additionally they managed to find the optimal pressure ratios for the reheating. A summary of the CI literature in coal power is given in Table 7.

A graphical depiction of current application of CI in coal industries discussed in this section is given in Figure 8.

Table 7. Summary of the CI literature in coal power

References	Method & Application	Analysis
Lv *et al.*, (2013)	SVM/Predict boiler emissions	SVM variant was enhanced with a combination of the least squares method and fuzzy membership grades to boost its performance.
Zhang *et al.*, (2014)	SVM/Predict ash content	Feature extraction was done using GA and SVM was applied to model the ash content.
Tan *et al.*, (2014), Zong *et al.*, 2013, Liu *et al.*, (2015)	ANN, HSA, GA/ Boiler simulation	Combination of modeling and optimization were applied to boiler systems.
Yang *et al.*, (2015b)	ANN, PSO, SVM/ Predict blockage in coal pipelines	CI applied to auxiliary parts of coal power plants.
Karimi *et al.*, (2014)	ANN/Predict ash fusion temperature	ANN was extensively trained on 1040 US coal samples from 12 states.
Dindarloo and Hower, (2015)	SVM/Predict Carbon content	SVM managed to predict carbon content with greater accuracy compared to other methods.
Wang *et al.*, (2014a)	DE/Improve power plant efficiency	Achieved boiler optimization.
Wang *et al.*, (2014b)	DE/Improve plant economics	Managed to decrease the cost of electricity by 2% for a single reheating unit.

GAS TURBINE PLANTS AND COMBINED CYCLE SYSTEMS

In gas power plants, the energy source is natural gas which acts as fuel to a gas turbine (GT) engine. Accurate fault diagnosis of the GT is critical in gas power plants to ensure no disruption in power generation. CI has been used in such plants to aid with the fault diagnosis. Having multiple components, GT generators are highly complex. Wong *et al.*, (2013) developed a real-time

Figure 8. Current applications of CI in the coal industry

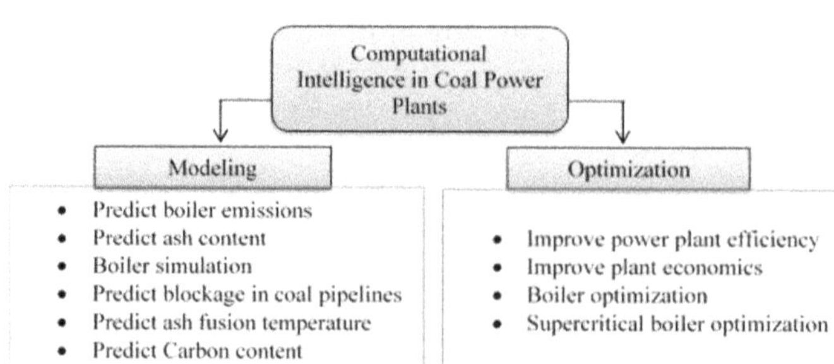

fault diagnosis system for GT generators using ANN (the extreme learning machine version). They used the neural n*et al*ong with other techniques like the wavelet packet transform, time-domain statistical methods and kernel principal component analysis. These techniques were applied for feature extraction and data processing. In their comparative studies they showed that their neural network performed much better in fault diagnosis as compared to the SVM approach.

Treating the GT system as an inverse problem, neural nets have been used to develop dynamic linear models. This was done in Yari *et al.*, (2013) on a V94.2 Siemens GT engine with the nominal power capacity of 162.1 MW and the nominal frequency of 50 Hz. In that work the following turbine parameters were identified as inputs and outputs to the model, given as follows:

- **Inputs:** Fuel pressure valve angle and inlet guide vane angle
- **Outputs:** Compressor output pressure, compressor output temperature, fuel pressure, turbine output power and turbine output temperature

In Yari *et al.*, (2013), it was stated that the ANN predicted the GT system dynamics to an accuracy of 96%. In addition to predictive modeling, some classification work has also been done on GT systems. For instance in Lilo *et al.*, (2015), the authors were interested in classifying the vibrations of the GT system - since high GT vibrations during operations could result to issues in safety, control and mechanical damage. They used an ANN to model the system's vibration and consequently classify components with high vibrations. An application of classification for fault detection in combined cycle plants could be seen in Berahman *et al.*, (2013). In that work, SVM was used in tandem with PCA to detect faults occurring in high pressure boiler drums in a combined cycle plant.

Gas power plants are often optimized by operating in a combined cycle setting. In such a setting the GT exhaust gas is supplied to a boiler which generates steam which drives a steam turbine. This way the thermal energy contained in the exhaust gas from the GT does not go to waste. In Rashid *et al.*, (2014), a combined cycle power plant was modeled using a feed-forward ANN enhanced by a PSO algorithm. Taking the ambient temperature, atmospheric pressure and relative humidity and exhaust vacuum as input parameters, the model predicts the average hourly output of the plant. A similar work was carried out by Fainti *et al.*, (2016) where a hierarchical method based ANN was employed to predict the output power of a combined cycle plant. Their work emphasized on the performance and effectiveness of the ANN when applied for this purpose.

Optimization efforts are also targeted towards control system optimization of gas as well as combined cycle plants. This can be observed in the work of Haji and Monje, (2017). In that work an enhanced PSO was developed to optimize the controller gains of a fuzzy-PID controller. This optimization is shown to significantly improve the performance of the combined cycle plant. The authors in Haji and Monje, (2017) rigorously compared their approach to other current optimization techniques: conventional PSO, Comprehensive Learning PSO (CLPSO), Heterogeneous CLPSO (HCLPSO), Genetic Algorithm (GA), Differential Evolution (DE), and Artificial Bee Colony (ABC) algorithm. Another application of control optimization in GT systems could be seen in Mansourabad *et al.*, (2013). In that work, a PSO-fuzzy PID controller was developed on a Matlab/Simulink platform. This work was focused on the start-up and operation of the GT - where the turbine speed and exhaust temperature is the controlled parameters. They measured the performance of the control strategy by measuring the features of the control response: rise time, settling time, overshoot and steady state error.

A similar work was carried out in Saika and Sahu (2013). The authors of that work used the firefly algorithm (FA) (a type of swarm intelligence technique) to optimize the control parameters in a combined cycle gas turbine power plant. Using the approach for frequency control, the authors optimized the PID controller gains in their plant. They ran a comparative study and sensitivity analysis on the optimized controller performance against other controllers. They found that the optimized PID controller was more robust. An interesting application of optimization techniques could be seen in the work of Ehyaei *et al.*, (2015). There the authors optimized the design of a combined cycle plant – while considering a fog inlet air cooling system. The fog inlet air cooling system is commonly employed to cool the air entering the GT. This type of air inlet cooling strategy increases the combustion efficiency of the GT. The authors of Ehyaei *et al.*, (2015) employed a GA for the optimization of the system. Their optimization efforts managed to increase the average power of the plant by 17.24% raising its overall efficiency by about 3.5%.

In Sadatsakkak *et al.*, (2015), the authors optimized both the thermal as well as economic aspects of an irreversible regenerative closed Brayton cycle. They considered the maximization of three objective functions: output power, ecological function and the thermo-economic criterion. The optimization was performed using the non-dominated sorting GA - II (NSGA-II) algorithm. Along the same line of research, the work of Bakhshmand *et al.*, (2015) involved the thermo-economic optimization of a combined cycle plant. Using GA, the optimal design parameter which minimizes the cost and maximizes the energy efficiency of the plant was obtained. As seen in Nadir *et al.*, (2016), PSO has also been used for thermo-economic optimization of combined cycle plants. In that work, the optimal design parameters for their heat recovery steam generators (HRSG) was obtained. The optimization they performed accounted for: the gas turbine outlet temperature, gas mass flow rate, number of HRSG levels and selling price of power. In steam turbine optimization, Guo and Zhang (2007) prove that data mining and fault diagnosis can be implemented using neural network and genetic algorithm with high precision. A summary of the CI literature in GT and Combined Cycle Systems is given in Table 8.

The recent key applications discussed in this section are presented graphically in Figure 9.

Table 8. Summary of the CI literature in GT and combined cycle systems

References	Method & Application	Analysis
Wong *et al.*, (2013)	ANN, SVM/Fault Diagnostics	Comparative studies showed that the ANN performed much better in fault diagnosis as compared to the SVM approach.
Yari *et al.*, (2013)	ANN/GT Simulation	Dynamic linear modelling using ANN which predicted the GT system dynamics to an accuracy of 96%.
Rashid *et al.*, (2014), Fainti *et al.*, (2016)	ANN, PSO/Plant Simulation	ANN for prediction of the output power of a combined cycle plant. The performance and effectiveness of the ANN was observed.
Lilo *et al.*, (2015)	ANN/GT Fault Diagnostics	ANN was used to model the system's vibration and consequently classify components with high vibrations.
Berahman *et al.*, (2013)	SVM/Boiler fault detection	SVM was used along with PCA to detect faults occurring in high-pressure boiler drums in a combined cycle plant.
Haji and Monje, (2017), Mansourabad *et al.*, (2013), Saika and Sahu (2013)	PSO variant, DE, GA, ABC, FA/Control Optimization	Rigorous comparison of the proposed approach (enhanced PSO) against the conventional PSO.
Ehyaei *et al.*, (2015)	GA/GT air inlet cooling	The GA strategy managed to increase the average power of the plant by 17.24%; raising its overall efficiency by about 3.5%.
Sadatsakkak *et al.*, (2015), Nadir *et al.*, (2016)	NSGA-II, GA, PSO/ Thermo-economic optimization	Technique successfully optimized the output power, ecological function and the thermo-economic criterion.
Guo and Zhang (2007)	ANN/ST Fault Diagnostics	ANN and genetic algorithm were used for data mining and fault diagnosis of steam turbine.

FINAL REMARKS

This chapter outlines the current applications of CI in energy generation systems. The CI techniques were generally divided into two classes: methods used for predictive modeling and techniques utilized for parametric optimization. These methods have been used broadly in engineering applications across the main energy generation industries. Many unexplored opportunities for implementation of CI in this sector remain. Additionally, hybrid methods that merge both classes of CI are becoming increasingly prominent. Such methods offer an all-in-one solution; combining prediction/modeling and optimization in a single system (Coban, 2014; Zhang *et al.*, 2014; Yang *et al.*, 2015b).

A potential avenue of research in this area is algorithmic enhancement via parameter tuning. All CI techniques require the tuning of its parameters and a suitable initial condition prior to algorithmic execution. The optimal settings of these features in the metaheuristic give it better customizability to the problem at hand. Such settings would significantly influence its performance. Research into methods for parametric tuning is definitely crucial for solving the complex problems of engineering. Besides works involving direct modeling and optimization, research could also be done on analyzing and testing the behavior of metaheuristics in various scenarios. Such studies would give future practitioners insight on selecting the suitable technique for solving a particular problem. This sort of data would considerably cut down implementation time and cost in the solution procedure.

Figure 9. Current implementations of CI in GT systems and combined cycle plants

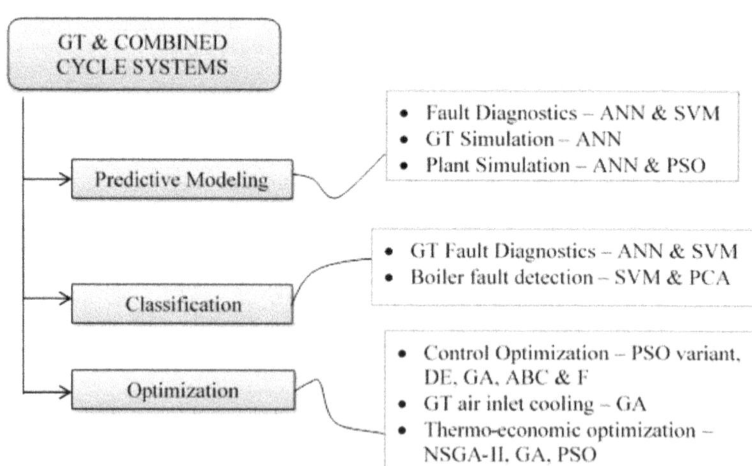

REFERENCES

Abdi, S., & Afshar, K. (2013). Application of IPSO-Monte Carlo for optimal distributed generation allocation and sizing. *International Journal of Electrical Power & Energy Systems*, *44*(1), 786–797. doi:10.1016/j.ijepes.2012.08.006

Aboutalebi, M., Bozorg Haddad, O., & Loáiciga, H. A. (2015). Optimal monthly reservoir operation rules for hydropower generation derived with SVR-NSGAII. *Journal of Water Resources Planning and Management*, *141*(11), 04015029. doi:10.1061/(ASCE)WR.1943-5452.0000553

Ahmed, J., & Salam, Z. (2014). A Maximum Power Point Tracking (MPPT) for PV system using Cuckoo Search with partial shading capability. *Applied Energy*, *119*, 118–130. doi:10.1016/j.apenergy.2013.12.062

Allegorico, C., & Mantini, V. (2014). A data-driven approach for on-line gas turbine combustion monitoring using classification models. In *Second European Conference of the Prognostics and Health Management Society* (pp. 92-100). Academic Press.

Alves, J. C. L., & Poppi, R. J. (2013). Biodiesel content determination in diesel fuel blends using near infrared (NIR) spectroscopy and support vector machines (SVM). *Talanta*, *104*, 155–161. doi:10.1016/j.talanta.2012.11.033 PMID:23597903

Ameli, A., Farrokhifard, M., Ahmadifar, A., & Haghifam, M. R. (2015). Distributed generation planning based on the distribution company's and the DG owner's profit maximization. *International Transactions on Electrical Energy Systems*, *25*(2), 216–232. doi:10.1002/etep.1835

Asadi, E., & Sadjadi, S. (2017). Optimization methods applied to renewable and sustainable energy: A review. *Uncertain Supply Chain Management*, *5*(1), 1–26. doi:10.5267/j.uscm.2016.6.001

Ata, R. (2015). Artificial neural networks applications in wind energy systems: A review. *Renewable & Sustainable Energy Reviews*, *49*, 534–562. doi:10.1016/j.rser.2015.04.166

Augusto, J. P. D. S. C., dos Santos Nicolau, A., & Schirru, R. (2015). PSO with dynamic topology and random keys method applied to nuclear reactor reload. *Progress in Nuclear Energy*, *83*, 191–196. doi:10.1016/j.pnucene.2015.03.009

Azizipour, M., Ghalenoei, V., Afshar, M. H., & Solis, S. S. (2016). Optimal operation of hydropower reservoir systems using weed optimization algorithm. *Water Resources Management*, *30*(11), 3995–4009. doi:10.100711269-016-1407-6

Ba, B. H., Prins, C., & Prodhon, C. (2016). Models for optimization and performance evaluation of biomass supply chains: An Operations Research perspective. *Renewable Energy*, *87*, 977–989. doi:10.1016/j.renene.2015.07.045

Babu, T. S., Rajasekar, N., & Sangeetha, K. (2015). Modified particle swarm optimization technique based maximum power point tracking for uniform and under partial shading condition. *Applied Soft Computing*, *34*, 613–624. doi:10.1016/j.asoc.2015.05.029

Bakhshmand, S. K., Saray, R. K., Bahlouli, K., Eftekhari, H., & Ebrahimi, A. (2015). Exergoeconomic analysis and optimization of a triple-pressure combined cycle plant using evolutionary algorithm. *Energy*, *93*, 555–567. doi:10.1016/j.energy.2015.09.073

Berahman, M., Safavi, A. A., & Shahrbabaki, M. R. (2013, December). Fault detection in Kerman combined cycle power plant boilers by means of support vector machine classifier algorithms and PCA. In *Control, Instrumentation, and Automation (ICCIA), 2013 3rd International Conference on* (pp. 290-295). IEEE. 10.1109/ICCIAutom.2013.6912851

Betiku, E., Okunsolawo, S. S., Ajala, S. O., & Odedele, O. S. (2015). Performance evaluation of artificial neural network coupled with generic algorithm and response surface methodology in modeling and optimization of biodiesel production process parameters from shea tree (Vitellaria paradoxa) nut butter. *Renewable Energy*, *76*, 408–417. doi:10.1016/j.renene.2014.11.049

Betiku, E., & Taiwo, A. E. (2015). Modeling and optimization of bioethanol production from breadfruit starch hydrolyzate vis-à-vis response surface methodology and artificial neural network. *Renewable Energy*, *74*, 87–94. doi:10.1016/j.renene.2014.07.054

Bhat, N. N., Dutta, S., Vashisth, T., Pal, S., Pal, S. K., & Sen, R. (2016). Tool condition monitoring by SVM classification of machined surface images in turning. *International Journal of Advanced Manufacturing Technology*, *83*(9-12), 1487–1502. doi:10.100700170-015-7441-3

Bhateshvar, Y. K., Mathur, H. D., Siguerdidjane, H., & Bhanot, S. (2015). Frequency stabilization for multi-area thermal–hydro power system using genetic algorithm-optimized fuzzy logic controller in deregulated environment. *Electric Power Components and Systems*, *43*(2), 146–156. doi:10.1080/15325008.2014.977462

Cao, H., Xin, Y., & Yuan, Q. (2016). Prediction of biochar yield from cattle manure pyrolysis via least squares support vector machine intelligent approach. *Bioresource Technology*, *202*, 158–164. doi:10.1016/j.biortech.2015.12.024 PMID:26708483

Casanova-Peláez, P. J., Palomar-Carnicero, J. M., Manzano-Agugliaro, F., & Cruz-Peragón, F. (2015). Olive cake improvement for bioenergy: The drying kinetics. *International Journal of Green Energy*, *12*(6), 559–569. doi:10.1080/15435075.2014.880347

Castillo-Villar, K. K. (2014). Metaheuristic algorithms applied to bioenergy supply chain problems: Theory, review, challenges, and future. *Energies*, *7*(11), 7640–7672. doi:10.3390/en7117640

Chang, H. C., Lin, S. C., Kuo, C. C., & Yu, H. P. (2014). Cloud monitoring for solar plants with support vector machine based fault detection system. *Mathematical Problems in Engineering*, 2014.

Chehouri, A., Younes, R., Ilinca, A., & Perron, J. (2015). Review of performance optimization techniques applied to wind turbines. *Applied Energy*, *142*, 361–388. doi:10.1016/j.apenergy.2014.12.043

Chen, J. H., Yau, H. T., & Hung, T. H. (2015). Design and implementation of FPGA-based Taguchi-chaos-PSO sun tracking systems. *Mechatronics*, *25*, 55–64. doi:10.1016/j.mechatronics.2014.12.004

Chen, K., & Yu, J. (2014). Short-term wind speed prediction using an unscented Kalman filter based state-space support vector regression approach. *Applied Energy*, *113*, 690–705. doi:10.1016/j.apenergy.2013.08.025

Chen, Y. H., Krishna, T., Emer, J. S., & Sze, V. (2017). Eyeriss: An energy-efficient reconfigurable accelerator for deep convolutional neural networks. *IEEE Journal of Solid-State Circuits*, *52*(1), 127–138. doi:10.1109/JSSC.2016.2616357

Cheng, C. T., Niu, W. J., Feng, Z. K., Shen, J. J., & Chau, K. W. (2015). Daily reservoir runoff forecasting method using artificial neural network based on quantum-behaved particle swarm optimization. *Water (Basel)*, *7*(8), 4232–4246. doi:10.3390/w7084232

Chitsaz, H., Amjady, N., & Zareipour, H. (2015). Wind power forecast using wavelet neural network trained by improved Clonal selection algorithm. *Energy Conversion and Management*, *89*, 588–598. doi:10.1016/j.enconman.2014.10.001

Chojaczyk, A. A., Teixeira, A. P., Neves, L. C., Cardoso, J. B., & Soares, C. G. (2015). Review and application of artificial neural networks models in reliability analysis of steel structures. *Structural Safety*, *52*, 78–89. doi:10.1016/j.strusafe.2014.09.002

Coban, R. (2014). Power level control of the TRIGA Mark-II research reactor using the multifeedback layer neural network and the particle swarm optimization. *Annals of Nuclear Energy*, *69*, 260–266. doi:10.1016/j.anucene.2014.02.019

Columbus, L. (2016). Internet of Things, Machine Learning & Robotics Are High Priorities For Developers In 2016. *Forbes*. Retrieved from https://www.forbes.com/sites/louiscolumbus/2016/06/18/internet-of-things-machine-learning-robotics-are-high-priorities-for-developers-in-2016/

Cong, T., Su, G., Qiu, S., & Tian, W. (2013). Applications of ANNs in flow and heat transfer problems in nuclear engineering: A review work. *Progress in Nuclear Energy*, *62*, 54–71. doi:10.1016/j.pnucene.2012.09.003

De Meyer, A., Cattrysse, D., Rasinmäki, J., & Van Orshoven, J. (2014). Methods to optimise the design and management of biomass-for-bioenergy supply chains: A review. *Renewable & Sustainable Energy Reviews*, *31*, 657–670. doi:10.1016/j.rser.2013.12.036

de Oliveira, M. V., & de Almeida, J. C. S. (2013). Application of artificial intelligence techniques in modeling and control of a nuclear power plant pressurizer system. *Progress in Nuclear Energy*, *63*, 71–85. doi:10.1016/j.pnucene.2012.11.005

de Souza, D. L., Granzotto, M. H., de Almeida, G. M., & Oliveira-Lopes, L. C. (2014). Fault detection and diagnosis using support vector machines-a SVC and SVR comparison. *Journal of Safety Engineering*, *3*(1), 18–29. doi:10.5923/j.safety.20140301.03

Deng, Y., Liu, Y., & Zhou, D. (2015). An improved genetic algorithm with initial population strategy for symmetric TSP. *Mathematical Problems in Engineering*.

Di Maio, F., Baronchelli, S., & Zio, E. (2014). Hierarchical differential evolution for minimal cut sets identification: Application to nuclear safety systems. *European Journal of Operational Research*, *238*(2), 645–652. doi:10.1016/j.ejor.2014.04.021

Diamantoulakis, P. D., Kapinas, V. M., & Karagiannidis, G. K. (2015). Big data analytics for dynamic energy management in smart grids. *Big Data Research*, *2*(3), 94–101. doi:10.1016/j.bdr.2015.03.003

Dindarloo, S. R., & Hower, J. C. (2015). Prediction of the unburned carbon content of fly ash in coal-fired power plants. *Coal Combustion and Gasification Products*, *7*(1), 19–29. doi:10.4177/CCGP-D-14-00009.1

Dong, Z. (2014). An artificial neural network compensated output feedback power-level control for modular high temperature gas-cooled reactors. *Energies*, *7*(3), 1149–1170. doi:10.3390/en7031149

Doucoure, B., Agbossou, K., & Cardenas, A. (2016). Time series prediction using artificial wavelet neural network and multi-resolution analysis: Application to wind speed data. *Renewable Energy*, *92*, 202–211. doi:10.1016/j.renene.2016.02.003

Ehyaei, M. A., Tahani, M., Ahmadi, P., & Esfandiari, M. (2015). Optimization of fog inlet air cooling system for combined cycle power plants using genetic algorithm. *Applied Thermal Engineering*, *76*, 449–461. doi:10.1016/j.applthermaleng.2014.11.032

Elkazaz, M. H., Hoballah, A. A., & Azmy, A. M. (2016). Operation optimization of distributed generation using artificial intelligent techniques. *Ain Shams Engineering Journal*, *7*(2), 855–866. doi:10.1016/j.asej.2016.01.008

Esfe, M. H., Saedodin, S., Sina, N., Afrand, M., & Rostami, S. (2015). Designing an artificial neural network to predict thermal conductivity and dynamic viscosity of ferromagnetic nanofluid. *International Communications in Heat and Mass Transfer*, *68*, 50–57. doi:10.1016/j.icheatmasstransfer.2015.06.013

Fainti, R., Nasiakou, A., Alamaniotis, M., & Tsoukalas, L. H. (2016). Hierarchical Method Based on Artificial Neural Networks for Power Output Prediction of a Combined Cycle Power Plant. *International Journal of Monitoring and Surveillance Technologies Research*, *4*(4), 20–32. doi:10.4018/IJMSTR.2016100102

Fan, F. H., Ma, Q., Ge, J., Peng, Q. Y., Riley, W. W., & Tang, S. Z. (2013). Prediction of texture characteristics from extrusion food surface images using a computer vision system and artificial neural networks. *Journal of Food Engineering*, *118*(4), 426–433. doi:10.1016/j.jfoodeng.2013.04.015

Fang, B., Yin, X., Tan, Y., Li, C., Gao, Y., Cao, Y., & Li, J. (2016). The contributions of cloud technologies to smart grid. *Renewable & Sustainable Energy Reviews*, *59*, 1326–1331. doi:10.1016/j.rser.2016.01.032

Fang, T., & Lahdelma, R. (2016). Optimization of combined heat and power production with heat storage based on sliding time window method. *Applied Energy*, *162*, 723–732. doi:10.1016/j.apenergy.2015.10.135

Farobie, O., Hasanah, N., & Matsumura, Y. (2015). Artificial neural network modeling to predict biodiesel production in supercritical methanol and ethanol using spiral reactor. *Procedia Environmental Sciences*, *28*, 214–223. doi:10.1016/j.proenv.2015.07.028

Fayek, H. M., Lech, M., & Cavedon, L. (2016, July). Modeling subjectiveness in emotion recognition with deep neural networks: Ensembles vs soft labels. In *Neural Networks (IJCNN), 2016 International Joint Conference on* (pp. 566-570). IEEE.

Feng, J., & Shen, W. Z. (2015). Solving the wind farm layout optimization problem using random search algorithm. *Renewable Energy*, *78*, 182–192. doi:10.1016/j.renene.2015.01.005

Ferrari, S., Lazzaroni, M., Piuri, V., Salman, A., Cristaldi, L., & Faifer, M. (2013). Computational intelligence models for solar radiation prediction. *2013 Instrumentation and Measurement Technology Conference (I2MTC)*.

Filgueiras, P. R., Alves, J. C. L., & Poppi, R. J. (2014). Quantification of animal fat biodiesel in soybean biodiesel and B20 diesel blends using near infrared spectroscopy and synergy interval support vector regression. *Talanta*, *119*, 582–589. doi:10.1016/j.talanta.2013.11.056 PMID:24401458

Fu, R., Harrison, R. F., King, S., & Mills, A. R. (2016, March). Lean burn combustion monitoring strategy based on data modelling. In *Aerospace Conference*, 2016 IEEE (pp. 1-11). IEEE. 10.1109/AERO.2016.7500876

Funes, E., Allouche, Y., Beltrán, G., & Jiménez, A. (2015). A Review: Artificial neural networks as tool for control food industry process. *Journal of Sensor Technology*, *5*(01), 28–43. doi:10.4236/jst.2015.51004

Gamerman, D., & Lopes, H. F. (2006). *Markov chain Monte Carlo: stochastic simulation for Bayesian inference*. Chapman and Hall/CRC.

Ganesan, T., Elamvazuthi, I., Shaari, K. Z. K., & Vasant, P. (2014). Multiobjective optimization of bioethanol production via hydrolysis using hopfield-enhanced differential evolution. In Contemporary advancements in information technology development in dynamic environments (pp. 340-359). doi:10.4018/978-1-4666-6252-0.ch017

Ganesan, T., Elamvazuthi, I., & Vasant, P. (2015). Multiobjective design optimization of a nano-CMOS voltage-controlled oscillator using game theoretic-differential evolution. *Applied Soft Computing*, *32*, 293–299. doi:10.1016/j.asoc.2015.03.016

Ganesan, T., Vasant, P., & Elamvazuthi, I. (2013). Hybrid neuro-swarm optimization approach for design of distributed generation power systems. *Neural Computing & Applications*, *23*(1), 105–117. doi:10.100700521-012-0976-4

Ganesan, T., Vasant, P., & Elamvazuthi, I. (2016). *Advances in Metaheuristics: Applications in Engineering Systems*. CRC Press. doi:10.1201/9781315297651

Gao, X., Yang, H., Lin, L., & Koo, P. (2015). Wind turbine layout optimization using multi-population genetic algorithm and a case study in Hong Kong offshore. *Journal of Wind Engineering and Industrial Aerodynamics*, *139*, 89–99. doi:10.1016/j.jweia.2015.01.018

García, J. A. M., & Mena, A. J. G. (2013). Optimal distributed generation location and size using a modified teaching–learning based optimization algorithm. *International Journal of Electrical Power & Energy Systems*, *50*, 65–75. doi:10.1016/j.ijepes.2013.02.023

Georgilakis, P. S., & Hatziargyriou, N. D. (2013). Optimal distributed generation placement in power distribution networks: Models, methods, and future research. *IEEE Transactions on Power Systems*, *28*(3), 3420–3428. doi:10.1109/TPWRS.2012.2237043

Ghimire, B. N., & Reddy, M. J. (2013). Optimal reservoir operation for hydropower production using particle swarm optimization and sustainability analysis of hydropower. *ISH Journal of Hydraulic Engineering*, *19*(3), 196–210. doi:10.1080/09715010.2013.796691

Glotić, A., & Zamuda, A. (2015). Short-term combined economic and emission hydrothermal optimization by surrogate differential evolution. *Applied Energy*, *141*, 42–56. doi:10.1016/j.apenergy.2014.12.020

Goyal, H., Hanmandlu, M., & Kothari, D. P. (2014). *An artificial intelligence based approach for control of small hydro power plants*. Centre for Energy Studies, Indian Institute of Technology.

Guo, Q., & Zhang, M. (2007). A novel approach for fault diagnosis of steam turbine based on neural network and genetic algorithm. *2008 International Joint Conference on Neural Networks*.

Haji, V. H., & Monje, C. A. (2017). Fractional order fuzzy-PID control of a combined cycle power plant using Particle Swarm Optimization algorithm with an improved dynamic parameters selection. *Applied Soft Computing*, *58*, 256–264. doi:10.1016/j.asoc.2017.04.033

Hammid, A. T., Sulaiman, M. H. B., & Abdalla, A. N. (2017). *Prediction of small hydropower plant power production in Himreen Lake dam (HLD) using artificial neural network*. Alexandria Engineering Journal.

He, Y., & Agarwal, R. K. (2014). Shape optimization of NREL S809 airfoil for wind turbine blades using a multiobjective genetic algorithm. *International Journal of Aerospace Engineering*.

Hejazi, H. A., Araghi, A. R., Vahidi, B., Hosseinian, S. H., Abedi, M., & Mohsenian-Rad, H. (2013). Independent distributed generation planning to profit both utility and DG investors. *IEEE Transactions on Power Systems*, *28*(2), 1170–1178. doi:10.1109/TPWRS.2012.2219325

Herman, M. R., Nejadhashemi, A. P., Daneshvar, F., Abouali, M., Ross, D. M., Woznicki, S. A., & Zhang, Z. (2016). Optimization of bioenergy crop selection and placement based on a stream health indicator using an evolutionary algorithm. *Journal of Environmental Management*, *181*, 413–424. doi:10.1016/j.jenvman.2016.07.005 PMID:27420165

Hernández-Travieso, J. G., Travieso, C. M., Alonso, J. B., & Dutta, M. K. (2014, August). Solar radiation modelling for the estimation of the solar energy generation. In *Contemporary Computing (IC3), 2014 Seventh International Conference on* (pp. 536-541). IEEE. 10.1109/IC3.2014.6897230

Hill, N. J., & Parks, G. T. (2015). Pressurized water reactor in-core nuclear fuel management by tabu search. *Annals of Nuclear Energy*, *75*, 64–71. doi:10.1016/j.anucene.2014.07.051

Hodzic, M., & Tai, L. C. (2016). Grey Predictor reference model for assisting particle swarm optimization for wind turbine control. *Renewable Energy*, *86*, 251–256. doi:10.1016/j.renene.2015.08.001

Hossain, M. A., & Ferdous, I. (2015). Autonomous robot path planning in dynamic environment using a new optimization technique inspired by bacterial foraging technique. *Robotics and Autonomous Systems*, *64*, 137–141. doi:10.1016/j.robot.2014.07.002

Hou, P., Hu, W., Soltani, M., & Chen, Z. (2015). Optimized placement of wind turbines in large-scale offshore wind farm using particle swarm optimization algorithm. *IEEE Transactions on Sustainable Energy*, *6*(4), 1272–1282. doi:10.1109/TSTE.2015.2429912

Huda, A. S. N., & Živanović, R. (2017). Large-scale integration of distributed generation into distribution networks: Study objectives, review of models and computational tools. *Renewable & Sustainable Energy Reviews*, *76*, 974–988. doi:10.1016/j.rser.2017.03.069

Ibrahim, W. Z. (2014). Particle Swarm Optimization to the U-tube steam generator in the nuclear power plant. *Nuclear Engineering and Design*, *280*, 94–98. doi:10.1016/j.nucengdes.2014.09.031

Iqbal, M., Azam, M., Naeem, M., Khwaja, A. S., & Anpalagan, A. (2014). Optimization classification, algorithms and tools for renewable energy: A review. *Renewable & Sustainable Energy Reviews*, *39*, 640–654. doi:10.1016/j.rser.2014.07.120

Jafarnejadsani, H., Pieper, J., & Ehlers, J. (2013). Adaptive control of a variable-speed variable-pitch wind turbine using radial-basis function neural network. *IEEE Transactions on Control Systems Technology*, *21*(6), 2264–2272. doi:10.1109/TCST.2012.2237518

Jamalipour, M., Sayareh, R., Gharib, M., Khoshahval, F., & Karimi, M. R. (2013). Quantum behaved particle swarm optimization with differential mutation operator applied to WWER-1000 in-core fuel management optimization. *Annals of Nuclear Energy*, *54*, 134–140. doi:10.1016/j.anucene.2012.11.008

Jayalal, M. L., Murty, S. S., & Baba, M. S. (2014). A survey of genetic algorithm applications in nuclear fuel management. *Journal of Nuclear Engineering and Technology*, *4*(1), 45–62.

Jha, S. K., Bilalovic, J., Jha, A., Patel, N., & Zhang, H. (2017). Renewable energy: Present research and future scope of Artificial Intelligence. *Renewable & Sustainable Energy Reviews*, *77*, 297–317. doi:10.1016/j.rser.2017.04.018

Jiang, B. T., Hines, J. W., & Zhao, F. Y. (2016b, June). Application of Fuzzy Weighted Support Vector Regression to Accidents Condition Monitoring of Nuclear Power Plant. In *2016 24th International Conference on Nuclear Engineering* (pp. V002T07A014-V002T07A014). American Society of Mechanical Engineers. 10.1115/ICONE24-60983

Jiang, P., Hu, Z., Liu, J., Yu, S., & Wu, F. (2016a). Fault diagnosis based on chemical sensor data with an active deep neural network. *Sensors (Basel)*, *16*(10), 1695. doi:10.339016101695 PMID:27754386

Juang, C. F., Chen, Y. H., & Jhan, Y. H. (2015). Wall-following control of a hexapod robot using a data-driven fuzzy controller learned through differential evolution. *IEEE Transactions on Industrial Electronics*, *62*(1), 611–619. doi:10.1109/TIE.2014.2319213

Kahla, S., Soufi, Y., Sedraoui, M., & Bechouat, M. (2015). On-Off control based particle swarm optimization for maximum power point tracking of wind turbine equipped by DFIG connected to the grid with energy storage. *International Journal of Hydrogen Energy*, *40*(39), 13749–13758. doi:10.1016/j.ijhydene.2015.05.007

Kaliannan, J., Baskaran, A., & Dey, N. (2015). Automatic generation control of Thermal-Thermal-Hydro power systems with PID controller using ant colony optimization. *International Journal of Service Science, Management, Engineering, and Technology*, *6*(2), 18–34. doi:10.4018/ijssmet.2015040102

Kansal, S., Kumar, V., & Tyagi, B. (2013). Optimal placement of different type of DG sources in distribution networks. *International Journal of Electrical Power & Energy Systems*, *53*, 752–760. doi:10.1016/j.ijepes.2013.05.040

Karabulut, Ş. (2015). Optimization of surface roughness and cutting force during AA7039/Al 2 O 3 metal matrix composites milling using neural networks and Taguchi method. *Measurement*, *66*, 139–149. doi:10.1016/j.measurement.2015.01.027

Karahroudi, M. R., Shirazi, S. M., & Sepanloo, K. (2013). Optimization of designing the core fuel loading pattern in a VVER-1000 nuclear power reactor using the genetic algorithm. *Annals of Nuclear Energy*, *57*, 142–150. doi:10.1016/j.anucene.2013.01.051

Karimi, S., Jorjani, E., Chehreh Chelgani, S., & Mesroghli, S. (2014). Multivariable Regression and Adaptive Neurofuzzy Inference System Predictions of Ash Fusion Temperatures Using Ash Chemical Composition of US Coals. *Journal of Fuels*.

Kashi, S., Minuchehr, A., Poursalehi, N., & Zolfaghari, A. (2014). Bat algorithm for the fuel arrangement optimization of reactor core. *Annals of Nuclear Energy*, *64*, 144–151. doi:10.1016/j.anucene.2013.09.044

Kazem, H. A., Yousif, J. H., & Chaichan, M. T. (2016). Modeling of daily solar energy system prediction using support vector machine for Oman. *International Journal of Applied Engineering Research*, *11*(20), 10166–10172.

Khademi, M., Moadel, M., & Khosravi, A. (2016). Power Prediction and Technoeconomic Analysis of a Solar PV Power Plant by MLP-ABC and COMFAR III, considering Cloudy Weather Conditions. *International Journal of Chemical Engineering*.

Khorasani, A., & Yazdi, M. R. S. (2015). Development of a dynamic surface roughness monitoring system based on artificial neural networks (ANN) in milling operation. *International Journal of Advanced Manufacturing Technology*, 1–11.

Kiruthiga, D., & Amudha, T. (2016). Optimal reservoir release for hydropower generation maximization using particle swarm optimization. In *Innovations in Bio-Inspired Computing and Applications* (pp. 577–585). Cham: Springer. doi:10.1007/978-3-319-28031-8_51

Kleshchyova, M. A., & Nakhabov, A. V. (2015). Prediction of electric power consumption by nuclear power plants for covering plant needs using data mining methods. *Nuclear Energy and Technology*, *1*(4), 243–247. doi:10.1016/j.nucet.2016.02.010

Kotsiantis, S.B., Zaharakis, I., & Pintelas, P. (2007). *Supervised machine learning: A review of classification techniques*. Academic Press.

Kowsalya, M. (2014). Optimal size and siting of multiple distributed generators in distribution system using bacterial foraging optimization. *Swarm and Evolutionary Computation*, *15*, 58–65. doi:10.1016/j.swevo.2013.12.001

Kulkarni, M. N. K., Patekar, M. S., Bhoskar, M. T., Kulkarni, M. O., Kakandikar, G. M., & Nandedkar, V. M. (2015). Particle swarm optimization applications to mechanical engineering-A review. *Materials Today: Proceedings*, *2*(4-5), 2631–2639.

Kumar, T., & Thakur, T. (2014, August). Comparative analysis of particle swarm optimization variants on distributed generation allocation for network loss minimization. In *Networks & Soft Computing (ICNSC), 2014 First International Conference on* (pp. 167-171). IEEE. 10.1109/CNSC.2014.6906682

Kusiak, A., Zhang, Z., & Li, M. (2010). Optimization of Wind Turbine Performance With Data-Driven Models. IEEE Transactions on Sustainable Energy, 1(2), 66-76. doi:10.1109/TSTE.2010.2046919

Langdon, W. B., & Harman, M. (2015). Optimizing existing software with genetic programming. *IEEE Transactions on Evolutionary Computation*, *19*(1), 118–135. doi:10.1109/TEVC.2013.2281544

Le Anh, T. N., Vo, D. N., Ongsakul, W., Vasant, P., & Ganesan, T. (2015). Cuckoo optimization algorithm for optimal power flow. In *Proceedings of the 18th Asia Pacific Symposium on Intelligent and Evolutionary Systems* (pp. 479-493). Springer. 10.1007/978-3-319-13359-1_37

Li, C., Yang, Z., Yan, H., & Wang, T. (2014, March). The application and research of the GA-BP neural network algorithm in the MBR membrane fouling. In *Abstract and Applied Analysis* (Vol. 2014). Hindawi Publishing Corporation.

Li, C., Zhou, J., Xiao, J., & Xiao, H. (2013, January). Vibration fault diagnosis of hydroelectric generating unit using gravitational search based kernel clustering method. *Zhongguo Dianji Gongcheng Xuebao*, *33*(2), 98–104.

Li, J., Meng, J., Kang, X., Long, Z., & Huang, X. (2017). Using Wireless Sensor Networks to Achieve Intelligent Monitoring for High-Temperature Gas-Cooled Reactor. *Science and Technology of Nuclear Installations*.

Lilo, M. A., Latiff, L. A., Abu, A. B. H., Al Mashhadany, Y. I., & Ilijan, A. K. (2015, May). Gas Turbine bearing and vibration classification of using multi-layer Neural Network. In *Smart Sensors and Application (ICSSA), 2015 International Conference on* (pp. 20-23). IEEE.

Liu, J., Seraoui, R., Vitelli, V., & Zio, E. (2013b). Nuclear power plant components condition monitoring by probabilistic support vector machine. *Annals of Nuclear Energy*, *56*, 23–33. doi:10.1016/j.anucene.2013.01.005

Liu, J., Vitelli, V., Seraoui, R., Di Maio, F., & Zio, E. (2013a, September). Short-Term Prediction for Nuclear Power Plant Failure Scenarios Using an Ensemble-based Approach. In ESREL 2013 (pp. 1-5). doi:10.1201/b15938-275

Liu, J., & Zio, E. (2015, July). Adaptive Support Vector Regression for Long-Term Prediction under Nonstationay Environments: Prognostics of Components in Nuclear Power Plants. *2015 International Conference on Quality, Reliability, Risk, Maintenance, and Safety Engineering (QR2MSE 2015)*.

Liu, L., Li, K., & Gao, J. (2015, December). Optimization of coal-fired boiler using neural network improved by genetic algorithm. In *Advanced Information Technology, Electronic and Automation Control Conference (IAEAC), 2015 IEEE* (pp. 567-571). IEEE. 10.1109/IAEAC.2015.7428617

Lv, Y., Liu, J., Yang, T., & Zeng, D. (2013). A novel least squares support vector machine ensemble model for NO x emission prediction of a coal-fired boiler. *Energy*, *55*, 319–329. doi:10.1016/j.energy.2013.02.062

Mahi, M., Baykan, Ö. K., & Kodaz, H. (2015). A new hybrid method based on particle swarm optimization, ant colony optimization and 3-opt algorithms for traveling salesman problem. *Applied Soft Computing*, *30*, 484–490. doi:10.1016/j.asoc.2015.01.068

Mahmoud, M. S., & Fouad, M. (2015). *Control and optimization of distributed generation systems*. Springer. doi:10.1007/978-3-319-16910-1

Mahmoudi, S. M., Aghaie, M., Bahonar, M., & Poursalehi, N. (2016). A novel optimization method, Gravitational Search Algorithm (GSA), for PWR core optimization. *Annals of Nuclear Energy*, *95*, 23–34. doi:10.1016/j.anucene.2016.04.035

Manjarres, D., Landa-Torres, I., Gil-Lopez, S., Del Ser, J., Bilbao, M. N., Salcedo-Sanz, S., & Geem, Z. W. (2013). A survey on applications of the harmony search algorithm. *Engineering Applications of Artificial Intelligence*, *26*(8), 1818–1831. doi:10.1016/j.engappai.2013.05.008

Mansourabad, A. M., Beheshti, M. T. H., & Simab, M. (2013). A Hybrid PSO Fuzzy PID controller for gas turbine speed control. *International Journal of Control and Automation*, *6*(2), 13–24.

Maran, J. P., & Priya, B. (2015). Comparison of response surface methodology and artificial neural network approach towards efficient ultrasound-assisted biodiesel production from muskmelon oil. *Ultrasonics Sonochemistry*, *23*, 192–200. doi:10.1016/j.ultsonch.2014.10.019 PMID:25457517

Mehdinejad, M., Mohammadi-Ivatloo, B., & Dadashzadeh-Bonab, R. (2017). Energy production cost minimization in a combined heat and power generation systems using cuckoo optimization algorithm. *Energy Efficiency*, *10*(1), 81–96. doi:10.100712053-016-9439-6

Mehra, M., Jayalal, M. L., Arul, A. J., Rajeswari, S., Kuriakose, K. K., & Murty, S. S. (2013). Study on different crossover mechanisms of genetic algorithm for test interval optimization for nuclear power plants. *International Journal of Intelligent Systems and Applications*, *6*(1), 20–28. doi:10.5815/ijisa.2014.01.03

Meng, X., Jia, M., & Wang, T. (2014). Neural network prediction of biodiesel kinematic viscosity at 313K. *Fuel*, *121*, 133–140. doi:10.1016/j.fuel.2013.12.029

Mikulandrić, R., Lončar, D., Böhning, D., Böhme, R., & Beckmann, M. (2014). Artificial neural network modelling approach for a biomass gasification process in fixed bed gasifiers. *Energy Conversion and Management*, *87*, 1210–1223. doi:10.1016/j.enconman.2014.03.036

Mirkouei, A., Haapala, K. R., Sessions, J., & Murthy, G. S. (2016). Evolutionary Optimization of bioenergy supply chain cost with uncertain forest biomass quality and availability. *Proceedings of the IIE/ISERC.*

Mirkouei, A., Haapala, K. R., Sessions, J., & Murthy, G. S. (2017). A review and future directions in techno-economic modeling and optimization of upstream forest biomass to bio-oil supply chains. *Renewable & Sustainable Energy Reviews*, *67*, 15–35. doi:10.1016/j.rser.2016.08.053

Moradi, M. H., Zeinalzadeh, A., Mohammadi, Y., & Abedini, M. (2014). An efficient hybrid method for solving the optimal sitting and sizing problem of DG and shunt capacitor banks simultaneously based on imperialist competitive algorithm and genetic algorithm. *International Journal of Electrical Power & Energy Systems*, *54*, 101–111. doi:10.1016/j.ijepes.2013.06.023

Moreno, S. R., & And Kaviski, E. (2015). Daily Scheduling Of Small Hydro Power Plants Dispatch With Modified Particles Swarm Optimization. *Pesquisa Operacional*, *35*(1), 25–37. doi:10.1590/0101-7438.2015.035.01.0025

Mortazavi, S. M., Soltani, M. R., & Motieyan, H. (2015). A Pareto optimal multi-objective optimization for a horizontal axis wind turbine blade airfoil sections utilizing exergy analysis and neural networks. *Journal of Wind Engineering and Industrial Aerodynamics*, *136*, 62–72. doi:10.1016/j.jweia.2014.10.009

Mortezazadeh, K., Norouzi, A., Zolfaghari, A., & Aghaie, M. (2015). Optimization of refueling cycle length by an enhanced PSO with novel mutation operator. *Progress in Nuclear Energy*, *78*, 251–257. doi:10.1016/j.pnucene.2014.09.018

Nabati, E. G., & Thoben, K. D. (2017). Big Data Analytics in the Maintenance of Off-Shore Wind Turbines: A Study on Data Characteristics. In Dynamics in Logistics (pp. 131-140). Springer International Publishing. doi:10.1007/978-3-319-45117-6_12

Naderi, B., Rahmani, S., & Rahmani, S. (2014). A Multiobjective Iterated Greedy Algorithm for Truck Scheduling in Cross-Dock Problems. *Journal of Industrial Engineering.*

Nadir, M., Ghenaiet, A., & Carcasci, C. (2016). Thermo-economic optimization of heat recovery steam generator for a range of gas turbine exhaust temperatures. *Applied Thermal Engineering*, *106*, 811–826. doi:10.1016/j.applthermaleng.2016.06.035

Oró, E., Depoorter, V., Garcia, A., & Salom, J. (2015). Energy efficiency and renewable energy integration in data centres. Strategies and modelling review. *Renewable & Sustainable Energy Reviews*, *42*, 429–445. doi:10.1016/j.rser.2014.10.035

Osório, G. J., Matias, J. C. O., & Catalão, J. P. S. (2015). Short-term wind power forecasting using adaptive neuro-fuzzy inference system combined with evolutionary particle swarm optimization, wavelet transform and mutual information. *Renewable Energy*, *75*, 301–307. doi:10.1016/j.renene.2014.09.058

Pappachen, A., & Fathima, A. P. (2016). Load frequency control in deregulated power system integrated with SMES–TCPS combination using ANFIS controller. *International Journal of Electrical Power & Energy Systems*, *82*, 519–534. doi:10.1016/j.ijepes.2016.04.032

Park, J., & Law, K. H. (2015). Layout optimization for maximizing wind farm power production using sequential convex programming. *Applied Energy*, *151*, 320–334. doi:10.1016/j.apenergy.2015.03.139

Paul, J., Joseph, T., & Sreedharan, S. 2013, March. PSO based generator rescheduling for relieving transmission overload. In Automation, Computing, Communication, Control and Compressed Sensing (iMac4s), 2013 International Multi-Conference on (pp. 409-414). IEEE. doi:10.1109/iMac4s.2013.6526446

Petković, D., Ćojbašić, Ž., Nikolić, V., Shamshirband, S., Kiah, M. L. M., Anuar, N. B., & Wahab, A. W. A. (2014). Adaptive neuro-fuzzy maximal power extraction of wind turbine with continuously variable transmission. *Energy*, *64*, 868–874. doi:10.1016/j.energy.2013.10.094

Pookpunt, S., & Ongsakul, W. (2013). Optimal placement of wind turbines within wind farm using binary particle swarm optimization with time-varying acceleration coefficients. *Renewable Energy*, *55*, 266–276. doi:10.1016/j.renene.2012.12.005

Poursalehi, N. (2015). Development of a new approach, evolutionary harmony search algorithm, for the LPO problem. *Progress in Nuclear Energy*, *81*, 78–90. doi:10.1016/j.pnucene.2015.01.004

Poursalehi, N., Zolfaghari, A., & Minuchehr, A. (2013b). Multi-objective loading pattern enhancement of PWR based on the discrete firefly algorithm. *Annals of Nuclear Energy*, *57*, 151–163. doi:10.1016/j.anucene.2013.01.043

Poursalehi, N., Zolfaghari, A., & Minuchehr, A. (2013c). Differential harmony search algorithm to optimize PWRs loading pattern. *Nuclear Engineering and Design*, *257*, 161–174. doi:10.1016/j.nucengdes.2013.01.020

Poursalehi, N., Zolfaghari, A., Minuchehr, A., & Moghaddam, H. K. (2013a). Continuous firefly algorithm applied to PWR core pattern enhancement. *Nuclear Engineering and Design*, *258*, 107–115. doi:10.1016/j.nucengdes.2013.02.011

Prakash, S., & Sinha, S. K. (2014). Simulation based neuro-fuzzy hybrid intelligent PI control approach in four-area load frequency control of interconnected power system. *Applied Soft Computing*, *23*, 152–164. doi:10.1016/j.asoc.2014.05.020

Prakash, S., & Sinha, S. K. (2015). Load frequency control of multi-area power systems using neuro-fuzzy hybrid intelligent controllers. *Journal of the Institution of Electronics and Telecommunication Engineers*, *61*(5), 526–532. doi:10.1080/03772063.2015.1024765

Precup, R. E., David, R. C., Petriu, E. M., Preitl, S., & Rădac, M. B. (2013). Fuzzy logic-based adaptive gravitational search algorithm for optimal tuning of fuzzy-controlled servo systems. *IET Control Theory & Applications*, *7*(1), 99–107. doi:10.1049/iet-cta.2012.0343

Puig-Arnavat, M., Hernández, J.A., Bruno, J.C., & Coronas, A. (2013). Artificial neural network models for biomass gasification in fluidized bed gasifiers. *Biomass and Bioenergy, 49*, 279-289.

Qazi, A., Fayaz, H., Wadi, A., Raj, R. G., Rahim, N. A., & Khan, W. A. (2015). The artificial neural network for solar radiation prediction and designing solar systems: A systematic literature review. *Journal of Cleaner Production*, *104*, 1–12. doi:10.1016/j.jclepro.2015.04.041

Qi, J., & Cai, J. (2014). Error Modeling And Compensation of 3D Scanning Robot System Based on PSO-RBFNN. *International Journal on Smart Sensing and Intelligent Systems*, *7*(2), 837–855. doi:10.21307/ijssis-2017-684

Rafiq, M. Y., Bugmann, G., & Easterbrook, D. J. (2001). Neural network design for engineering applications. *Computers & Structures*, *79*(17), 1541–1552. doi:10.1016/S0045-7949(01)00039-6

Rajalakshmi, J., & Durairaj, S. (2016, February). Review on optimal distributed generation placement using particle swarm optimization algorithms. In *Emerging Trends in Engineering, Technology and Science (ICETETS), International Conference on* (pp. 1-6). IEEE. 10.1109/ICETETS.2016.7603088

Ramasamy, P., Chandel, S. S., & Yadav, A. K. (2015). Wind speed prediction in the mountainous region of India using an artificial neural network model. *Renewable Energy*, *80*, 338–347. doi:10.1016/j.renene.2015.02.034

Ranjani, M., & Murugesan, P. (2015). Optimal fuzzy controller parameters using PSO for speed control of Quasi-Z Source DC/DC converter fed drive. *Applied Soft Computing*, *27*, 332–356. doi:10.1016/j.asoc.2014.11.007

Rao, R. S., Ravindra, K., Satish, K., & Narasimham, S. V. L. (2013). Power loss minimization in distribution system using network reconfiguration in the presence of distributed generation. *IEEE Transactions on Power Systems*, *28*(1), 317–325. doi:10.1109/TPWRS.2012.2197227

Rashid, M., Kamal, K., Zafar, T., Sheikh, Z., Shah, A., & Mathavan, S. (2015, December). Energy prediction of a combined cycle power plant using a particle swarm optimization trained feedforward neural network. In *Mechanical Engineering, Automation and Control Systems (MEACS), 2015 International Conference on* (pp. 1-5). IEEE. 10.1109/MEACS.2015.7414935

Ren, Y., Suganthan, P. N., & Srikanth, N. (2016). A novel empirical mode decomposition with support vector regression for wind speed forecasting. *IEEE Transactions on Neural Networks and Learning Systems*, *27*(8), 1793–1798. doi:10.1109/TNNLS.2014.2351391 PMID:25222957

Saberian, A., Hizam, H., Radzi, M. A. M., Ab Kadir, M. Z. A., & Mirzaei, M. (2014). Modelling and prediction of photovoltaic power output using artificial neural networks. *International Journal of Photoenergy*.

Sacco, W. F., de Moura Meneses, A. A., & Henderson, N. (2013). Some studies on differential evolution variants for application to nuclear reactor core design. *Progress in Nuclear Energy*, *63*, 49–56. doi:10.1016/j.pnucene.2012.10.003

Sacco, W. F., & Henderson, N. (2014). Differential evolution with topographical mutation applied to nuclear reactor core design. *Progress in Nuclear Energy*, *70*, 140–148. doi:10.1016/j.pnucene.2013.09.012

Sacco, W. F., & Henderson, N. (2015). Balancing exploration and exploitation in differential evolution via variable scaling factors: An application to practical problems. *Progress in Nuclear Energy*, *83*, 365–373. doi:10.1016/j.pnucene.2015.04.014

Sacco, W. F., Henderson, N., & Rios-Coelho, A. C. (2014). Topographical clearing differential evolution: A new method to solve multimodal optimization problems. *Progress in Nuclear Energy*, *71*, 269–278. doi:10.1016/j.pnucene.2013.12.011

Sadatsakkak, S. A., Ahmadi, M. H., & Ahmadi, M. A. (2015). Thermodynamic and thermo-economic analysis and optimization of an irreversible regenerative closed Brayton cycle. *Energy Conversion and Management*, *94*, 124–129. doi:10.1016/j.enconman.2015.01.040

Sadeghi Mobarakeh, A., Rajabi-Ghahnavieh, A., & Haghighat, H. (2016). A bi-level approach for optimal contract pricing of independent dispatchable DG units in distribution networks. *International Transactions on Electrical Energy Systems*, *26*(8), 1685–1704. doi:10.1002/etep.2172

Saikia, L. C., & Sahu, S. K. (2013). Automatic generation control of a combined cycle gas turbine plant with classical controllers using firefly algorithm. *International Journal of Electrical Power & Energy Systems*, *53*, 27–33. doi:10.1016/j.ijepes.2013.04.007

Salcedo-Sanz, S., Rojo-Álvarez, J. L., Martínez-Ramón, M., & Camps-Valls, G. (2014). Support vector machines in engineering: An overview. *Wiley Interdisciplinary Reviews. Data Mining and Knowledge Discovery*, *4*(3), 234–267. doi:10.1002/widm.1125

Saleem, A., Taha, B., Tutunji, T., & Al-Qaisia, A. (2015). Identification and cascade control of servo-pneumatic system using Particle Swarm Optimization. *Simulation Modelling Practice and Theory*, *52*, 164–179. doi:10.1016/j.simpat.2015.01.007

Santamaría-Bonfil, G., Reyes-Ballesteros, A., & Gershenson, C. (2016). Wind speed forecasting for wind farms: A method based on support vector regression. *Renewable Energy*, *85*, 790–809. doi:10.1016/j.renene.2015.07.004

Santhiya, M., & Pappa, N. (2014, May). Evolutionary Algorithms based Fuzzy Logic Controller for Pressurized Water Nuclear Reactor. In *Advanced Communication Control and Computing Technologies (ICACCCT), 2014 International Conference on* (pp. 198-204). IEEE. 10.1109/ICACCCT.2014.7019428

Santoso, F., Garratt, M. A., & Anavatti, S. G. (2017). State-of-the-Art Intelligent Flight Control Systems in Unmanned Aerial Vehicles. *IEEE Transactions on Automation Science and Engineering*.

Sauhats, A., Petrichenko, R., Broka, Z., Baltputnis, K., & Sobolevskis, D. (2016, October). ANN-based forecasting of hydropower reservoir inflow. In *Power and Electrical Engineering of Riga Technical University (RTUCON), 2016 57th International Scientific Conference on* (pp. 1-6). IEEE.

Schlünz, E. B., Bokov, P. M., & van Vuuren, J. H. (2015). On multiobjective optimisation approaches for in-core fuel management optimisation. *Proceedings of the Topical Meeting on the Advances in Nuclear Fuel Management V (ANFM 2015)*.

Sebayang, A. H., Masjuki, H. H., Ong, H. C., Dharma, S., Silitonga, A. S., Kusumo, F., & Milano, J. (2017). Optimization of bioethanol production from sorghum grains using artificial neural networks integrated with ant colony. *Industrial Crops and Products*, *97*, 146–155. doi:10.1016/j.indcrop.2016.11.064

Senkerik, R., Zelinka, I., Pluhacek, M., Davendra, D., & Oplatková Kominkova, Z. (2014). Chaos enhanced differential evolution in the task of evolutionary control of selected set of discrete chaotic systems. *The Scientific World Journal*. PMID:25243230

Senvar, O., Turanoglu, E., & Kahraman, C. (2013). Usage of metaheuristics in engineering: A literature review. In *Meta-heuristics optimization algorithms in engineering, business, economics, and finance* (pp. 484–528). IGI Global. doi:10.4018/978-1-4666-2086-5.ch016

Sewsynker-Sukai, Y., Faloye, F., & Kana, E. B. G. (2017). Artificial neural networks: An efficient tool for modelling and optimization of biofuel production (a mini review). *Biotechnology, Biotechnological Equipment, 31*(2), 221–235. doi:10.1080/13102818.2016.1269616

Seyedmahmoudian, M., Rahmani, R., Mekhilef, S., Oo, A.M.T., Stojcevski, A., Soon, T.K., & Ghandhari, A.S. (2015). Simulation and hardware implementation of new maximum power point tracking technique for partially shaded PV system using hybrid DEPSO method. *IEEE Transactions on Sustainable Energy, 6*(3), 850-862.

Shamshirband, S., Tabatabaei, M., Aghbashlo, M., Yee, L., & Petković, D. (2016). Support vector machine-based exergetic modelling of a DI diesel engine running on biodiesel–diesel blends containing expanded polystyrene. *Applied Thermal Engineering, 94*, 727–747. doi:10.1016/j.applthermaleng.2015.10.140

Sheikhan, M., Shahnazi, R., & Yousefi, A. N. (2013). An optimal fuzzy PI controller to capture the maximum power for variable-speed wind turbines. *Neural Computing & Applications, 23*(5), 1359–1368. doi:10.100700521-012-1081-4

Sheng, S. (2015). *Improving component reliability through performance and condition monitoring data analysis*. Houston, TX: Wind Farm Data Management & Analysis North America.

Shirneshan, A., Samani, B. H., & Ghobadian, B. (2016). Optimization of biodiesel percentage in fuel mixture and engine operating conditions for diesel engine performance and emission characteristics by Artificial Bees Colony Algorithm. *Fuel, 184*, 518–526. doi:10.1016/j.fuel.2016.06.117

Shivaie, M., Kazemi, M. G., & Ameli, M. T. (2015). A modified harmony search algorithm for solving load-frequency control of non-linear interconnected hydrothermal power systems. *Sustainable Energy Technologies and Assessments, 10*, 53–62. doi:10.1016/j.seta.2015.02.001

Shui-jun, Y., Feng-cheng, X., Bo-yu, J., & Peng-fei, Z. (2012). Influence study of organic and inorganic additive to coal combustion characteristic. *Procedia Environmental Sciences, 12*, 459–467. doi:10.1016/j.proenv.2012.01.304

Shyam, R., HB, B.G., Kumar, S., Poornachandran, P., & Soman, K.P. (2015). Apache spark a big data analytics platform for smart grid. *Procedia Technology, 21*, 171–178. doi:10.1016/j.protcy.2015.10.085

Silveira, C. L., Belledeli, B. D., Soares, M., Treichel, H., & Mazutti, M. A. (2014). Use of artificial neural network for industrial prediction of final temperature of chicken carcasses. *Journal of Food Process Engineering*, *37*(1), 1–9. doi:10.1111/jfpe.12053

Simmhan, Y., Aman, S., Kumbhare, A., Liu, R., Stevens, S., Zhou, Q., & Prasanna, V. (2013). Cloud-based software platform for big data analytics in smart grids. *Computing in Science & Engineering*, *15*(4), 38–47. doi:10.1109/MCSE.2013.39

Singh, B., Pal, C., Mukherjee, V., Tiwari, P., Yadav, M. K., & Member, I. S. (2016). Distributed generation planning from power system performances viewpoints: A taxonomical survey. *Renewable & Sustainable Energy Reviews*.

Singh, K., Vasant, P., Elamvazuthi, I., & Kannan, R. (2015). PID tuning of servo motor using bat algorithm. *Procedia Computer Science*, *60*, 1798–1808. doi:10.1016/j.procs.2015.08.290

Singh, V. P., Vijay, V., Bhatt, M. S., & Chaturvedi, D. K. (2013, November). Generalized neural network methodology for short term solar power forecasting. In *Environment and Electrical Engineering (EEEIC), 2013 13th International Conference on* (pp. 58-62). IEEE. 10.1109/EEEIC-2.2013.6737883

Sloss, L. L. (2014). *Blending of coals to meet power station requirements*. IEA Clean Coal Centre.

Soomro, W. A., Elamvazuthi, I., Khan, A., Muralidharan, S., & Amudha, M. (2013). PID Controller Optimization using Artificial Fish Swarm Algorithm. *International Journal of Electrical and Electronics Research*, *1*, 11–18.

Syan, S., & Biswal, G. R. (2015). Frequency Control of an Isolated Hydro Power Plant Using Artificial Intelligence. *Computational Intelligence: Theories, Applications and Future Directions (WCI), 2015 IEEE Workshop*. 10.1109/WCI.2015.7495537

Tan, P., Xia, J., Zhang, C., Fang, Q., & Chen, G. (2014). Modeling and optimization of NOx emission in a coal-fired power plant using advanced machine learning methods. *Energy Procedia*, *61*, 377–380. doi:10.1016/j.egypro.2014.11.1129

Tayebiyan, A., Ali, T. A. M., Ghazali, A. H., & Malek, M. A. (2016). Optimization of exclusive release policies for hydropower reservoir operation by using genetic algorithm. *Water Resources Management*, *30*(3), 1203–1216. doi:10.100711269-015-1221-6

Vadivel, M., Karthigaivel, R., & Selvakumaran, S. (2014). Control of PWR Nuclear Reactor Power using Soft Computing Techniques based Proportional-Integral-Derivative Design. *International Journal of Applied Engineering Research*, *9*(24), 30605–30619.

Verma, S., Saha, S., & Mukherjee, V. (2016a). Optimal rescheduling of real power generation for congestion management using teaching-learning-based optimization algorithm. *Journal of Electrical Systems and Information Technology*.

Verma, T., Tiwana, A. P. S., Reddy, C. C., Arora, V., & Devanand, P. (2016b, January). Data Analysis to Generate Models Based on Neural Network and Regression for Solar Power Generation Forecasting. In *Intelligent Systems, Modelling and Simulation (ISMS), 2016 7th International Conference on* (pp. 97-100). IEEE. 10.1109/ISMS.2016.65

Viharos, Z.J., Sidló, C.I., Benczúr, A., Csempesz, J., Kis, K.B., Petrás, I., & Garzó, A. (2013). *"Big Data" Initiative as an IT Solution for Improved Operation and Maintenance of Wind Turbines*. Academic Press.

Vyas, S., Kumar, R., & Kavasseri, R. (2017). Data analytics and computational methods for anti-islanding of renewable energy based Distributed Generators in power grids. *Renewable & Sustainable Energy Reviews*, *69*, 493–502. doi:10.1016/j.rser.2016.11.116

Wan, C., Zhao, J., Song, Y., Xu, Z., Lin, J., & Hu, Z. (2015). Photovoltaic and solar power forecasting for smart grid energy management. *CSEE Journal of Power and Energy Systems*, *1*(4), 38–46. doi:10.17775/CSEEJPES.2015.00046

Wang, F., Zhen, Z., Mi, Z., Sun, H., Su, S., & Yang, G. (2015). Solar irradiance feature extraction and support vector machines based weather status pattern recognition model for short-term photovoltaic power forecasting. *Energy and Building*, *86*, 427–438. doi:10.1016/j.enbuild.2014.10.002

Wang, L., Yang, Y., Dong, C., Morosuk, T., & Tsatsaronis, G. (2014a). Multi-objective optimization of coal-fired power plants using differential evolution. *Applied Energy*, *115*, 254–264. doi:10.1016/j.apenergy.2013.11.005

Wang, L., Yang, Y., Dong, C., Morosuk, T., & Tsatsaronis, G. (2014b). Parametric optimization of supercritical coal-fired power plants by MINLP and differential evolution. *Energy Conversion and Management*, *85*, 828–838. doi:10.1016/j.enconman.2014.01.006

Wang, S., Zhang, N., Wu, L., & Wang, Y. (2016). Wind speed forecasting based on the hybrid ensemble empirical mode decomposition and GA-BP neural network method. *Renewable Energy*, *94*, 629–636. doi:10.1016/j.renene.2016.03.103

Watetakarn, S., & Premrudeepreechacharn, S. (2015, November). Forecasting of solar irradiance for solar power plants by artificial neural network. In Innovative Smart Grid Technologies-Asia (ISGT ASIA), 2015 IEEE (pp. 1-5). IEEE. doi:10.1109/ISGT-Asia.2015.7387180

Wen, S., Lan, H., Fu, Q., David, C. Y., & Zhang, L. (2015). Economic allocation for energy storage system considering wind power distribution. *IEEE Transactions on Power Systems*, *30*(2), 644–652. doi:10.1109/TPWRS.2014.2337936

Whiteman, J. K., & Kana, E. G. (2014). Comparative assessment of the artificial neural network and response surface modelling efficiencies for biohydrogen production on sugar cane molasses. *BioEnergy Research*, *7*(1), 295–305. doi:10.100712155-013-9375-7

Witten, I. H., Frank, E., Hall, M. A., & Pal, C. J. (2016). *Data Mining: Practical machine learning tools and techniques*. Morgan Kaufmann.

Wong, P. K., Wong, K. I., Vong, C. M., & Cheung, C. S. (2015). Modeling and optimization of biodiesel engine performance using kernel-based extreme learning machine and cuckoo search. *Renewable Energy*, *74*, 640–647. doi:10.1016/j.renene.2014.08.075

Wong, P. K., Yang, Z., Vong, C. M., & Zhong, J. (2014). Real-time fault diagnosis for gas turbine generator systems using extreme learning machine. *Neurocomputing*, *128*, 249–257. doi:10.1016/j.neucom.2013.03.059

Xie, D., & Bai, L. (2015, December). A hierarchical deep neural network for fault diagnosis on Tennessee-Eastman process. In *Machine Learning and Applications (ICMLA), 2015 IEEE 14th International Conference on* (pp. 745-748). IEEE. 10.1109/ICMLA.2015.208

Xiong, N., Molina, D., Ortiz, M. L., & Herrera, F. (2015). A walk into metaheuristics for engineering optimization: Principles, methods and recent trends. *International Journal Of CI Systems*, *8*(4), 606–636.

Yan, W. (2016, July). One-class extreme learning machines for gas turbine combustor anomaly detection. In *Neural Networks (IJCNN), 2016 International Joint Conference on* (pp. 2909-2914). IEEE. 10.1109/IJCNN.2016.7727567

Yang, T., Gao, X., Sellars, S. L., & Sorooshian, S. (2015a). Improving the multi-objective evolutionary optimization algorithm for hydropower reservoir operations in the California Oroville–Thermalito complex. *Environmental Modelling & Software*, *69*, 262–279. doi:10.1016/j.envsoft.2014.11.016

Yang, X. C., Yan, X. R., & Song, C. F. (2015b). Pressure prediction of coal slurry transportation pipeline based on particle swarm optimization kernel function extreme learning machine. *Mathematical Problems in Engineering*.

Yang, X. S. (2010). *Nature-inspired metaheuristic algorithms*. Luniver Press.

Yang, X. S., Bekdaş, G., & Nigdeli, S. M. (2016). Review and applications of metaheuristic algorithms in civil engineering. In *Metaheuristics and Optimization in Civil Engineering* (pp. 1–24). Springer International Publishing. doi:10.1007/978-3-319-26245-1_1

Yari, M., Shoorehdeli, M. A., & Yousefi, I. (2013, February). V94. 2 gas turbine identification using neural network. In *Robotics and Mechatronics (ICRoM), 2013 First RSI/ISM International Conference on* (pp. 523-529). IEEE.

Yu, R., Li, X., Tao, M., & Ke, Z. (2016, June). Fault Diagnosis of Feedwater Pump in Nuclear Power Plants Using Parameter-Optimized Support Vector Machine. In *2016 24th International Conference on Nuclear Engineering* (pp. V001T03A013-V001T03A013). American Society of Mechanical Engineers. 10.1115/ICONE24-60334

Yue, D., You, F., & Snyder, S. W. (2014). Biomass-to-bioenergy and biofuel supply chain optimization: Overview, key issues and challenges. *Computers & Chemical Engineering*, *66*, 36–56. doi:10.1016/j.compchemeng.2013.11.016

Zahedinejad, E., Schirmer, S., & Sanders, B. C. (2014). Evolutionary algorithms for hard quantum control. *Physical Review A.*, *90*(3), 032310. doi:10.1103/PhysRevA.90.032310

Zeng, J., & Qiao, W. (2013). Short-term solar power prediction using a support vector machine. *Renewable Energy*, *52*, 118–127. doi:10.1016/j.renene.2012.10.009

Zhang, Q., Ogren, R. M., & Kong, S. C. (2016, March 1). 2016b, A comparative study of biodiesel engine performance optimization using enhanced hybrid PSO–GA and basic GA. *Applied Energy*, *165*, 676–684. doi:10.1016/j.apenergy.2015.12.044

Zhang, X., Yu, X., & Qin, H. (2016c). Optimal operation of multi-reservoir hydropower systems using enhanced comprehensive learning particle swarm optimization. *Journal of Hydro-environment Research*, *10*, 50–63. doi:10.1016/j.jher.2015.06.003

Zhang, Y., Dong, Z., Liu, A., Wang, S., Ji, G., Zhang, Z., & Yang, J. (2015). Magnetic resonance brain image classification via stationary wavelet transform and generalized eigenvalue proximal support vector machine. *Journal of Medical Imaging and Health Informatics*, *5*(7), 1395–1403. doi:10.1166/jmihi.2015.1542

Zhang, Y., Wang, X., Zhuo, S., & Zhang, Y. (2016a). Pre-feasibility of building cooling heating and power system with thermal energy storage considering energy supply–demand mismatch. *Applied Energy*, *167*, 125–134. doi:10.1016/j.apenergy.2016.01.040

Zhang, Z., Yang, J., Wang, Y., Dou, D., & Xia, W. (2014). Ash content prediction of coarse coal by image analysis and GA-SVM. *Powder Technology*, *268*, 429–435. doi:10.1016/j.powtec.2014.08.044

Zong, W., Huang, G. B., & Chen, Y. (2013). Weighted extreme learning machine for imbalance learning. *Neurocomputing*, *101*, 229–242. doi:10.1016/j.neucom.2012.08.010

Chapter 2
Optimization of a Solar-powered Irrigation System

ABSTRACT

Optimization is now a crucial element in industrial applications involving sustainable alternative energy systems. During the design of such systems, the engineer/decision maker would often encounter noise factors when their system interacts with the environment (e.g., solar insolation and ambient temperature fluctuations). In this chapter, the sizing and design optimization of the solar powered irrigation system is considered. This problem is multivariate, noisy, nonlinear, and multiobjective (MO). This chapter is divided into two parts where two situations are considered during the optimization of the solar powered irrigation system. Part 1 is the MO design optimization of the mentioned system under constant weather conditions. Part 2 involves optimizing a more general form of the design problem by accounting for varying weather conditions, insolation, and ambient temperature. The details of the optimization procedures of the two cases are presented and discussed in this chapter.

DOI: 10.4018/978-1-7998-1710-9.ch002

PART 1: DESIGN OF SOLAR POWERED IRRIGATION SYSTEM UNDER CONSTANT WEATHER CONDITIONS

Overview

Currently, optimization problems are frequently encountered by engineers and scientists working on systems related to alternative energy and sustainable technologies (Elamvazuthi *et al.* 2011; Ganesan *et al.* 2012). Standard irrigation systems are usually powered by diesel generators or other fossil-fuel based power sources. However, issues such as the diminishing of fossil fuel resources and stricter environmental regulations have caused a surge in efforts to search for cost effective, eco-friendly and efficient alternative power sources. Hence, the idea of the utilization of solar energy to supply power to irrigation pumps has recently surfaced (Helikson *et al.*, 1991; Wong & Sumathy, 2001). In solar powered irrigation systems, design and sizing plays a critical role in reducing greenhouse emissions while ensuring the systems reliability and efficiency (Al-Ali *et al.*, 2001). Hence, this problem enters into the optimization domain where optimal design and sizing is done in a way that issues like cost saving, emissions, system efficiency as well power output are ensured or improved (Shivrath *et al.*, 2012; Carroqino *et al.*,2009). Optimization of solar powered irrigation systems has been carried out with various metaheuristic techniques such as: genetic algorithms (GA) (Holland, 1992) and particle swarm optimization (PSO) (Carroqino *et al.*,2009; Gouws & Lukhwareni, 2012). However when a system is designed such that it respects multiple aims, the need to carry out multi-objective (MO) optimization arises (Chen *et al.*, 1995).

In MO optimization, one approach that has been effectively used to measure the quality of solution sets (which construct the Pareto-frontier) is the Hypervolume Indicator (HVI) (Zitzler & Thiele, 1998). The HVI is specifically useful in cases where the Pareto frontier is unknown. Recently, this indicator has been frequently applied in many works involving MO problems (Beume *et al.*, 2007; Igel *et al.*, 2007; Knowles & Corne, 2003). The HVI is the only indicator which is strictly Pareto-compliant that can be used to measure the quality of solution sets (degree of dominance) in MO optimization problems (Beume *et al.*, 2007; Zitzler & Thiele, 1999).

The main approach presented in this work involves an evolutionary symbolic regression approach called Analytical Programming (AP) (Zelinka, 2002; Zelinka & Oplatkova, 2003). AP algorithm fuses the concepts of evolution and symbolic regression to form an efficient metaheuristic technique. Using concepts from Genetic Programming (GP) and Grammatical Evolution (GE), this generalized form of symbolic regression has been successfully applied in many case studies. For further insights on these techniques please refer to the works of Varacha, (2011), Oplatkova & Zelinka, (2007) and Oplatkova & Zelinka, (2006).

The central aim in this chapter is to solve and obtain a set of Pareto-efficient solution options for the MO optimization of the solar powered irrigation system. The solar powered irrigation system problem was developed and rigorously validated in Chen *et al*, (1995). The key ideas and details of the solution method in this chapter can be found in the author's past works (Ganesan *et al*., 2013 and Ganesan *et al*., 2017) solution method proposed in here is the hypervolume-driven symbolic regression approach (Hyp-AP). This method uses the HVI to drive the AP approach in search for dominant solutions in the objective space.

PROBLEM FORMULATION

In the solar powered irrigation system, the design parameters of the irrigation system influences the system's performance, efficiency and cost. In Chen *et al.* 1995, the objectives were set as the pump load/power output, f_1 (kW), overall efficiency, f_2 (%) and the fiscal savings, f_3 (USD). The design variables were: the maximum pressure, x_a (MPa), maximum temperature, x_b (K), maximum solar collector temperature, x_c (K), the fluid flowrate, x_d (kg/s), ambient temperature, Z_a (K) and the level of insolation, Z_b (K). The objective functions and the constraints of the decision variables are shown as follows:

$$f_1 = -(24.947 + 16.011x_d + 1.306x_b + 0.820x_b x_d - 0.785Z_a - 0.497x_d Z_a + 0.228x_a x_b + 0.212x_a - 0.15x_b^2 + 0.13x_a x_d - 0.11x_a^2 - 0.034x_b Z_a + 0.002x_a Z_a)10^{-3.24} \quad (2.1)$$

$$f_2 = -43.4783(0.18507 + 0.01041x_a + 0.0038Z_b - 0.00366Z_a - 0.0035x_c - 0.00157x_b) \quad (2.2)$$

$$f_3 = -(174695.73 + 112114.69x_d + 9133.8x_b + 5733.05x_b x_d - 5487.76Z_a - 3478.84x_d Z_a + 1586.48x_a x_b + 1486.84x_a - 1067.42x_b^{\ 2} + 916.26x_a x_d - 768.9x_a^2 - 242.88x_b Z_a + 152.4x_a Z_a)10^{-3.23} \tag{2.3}$$

$$0.3 \le x_a \le 3$$

$$450 \le x_b \le 520$$

$$520 \le x_c \le 800$$

$$0.01 \le x_d \le 0.2$$

$$293 \le Z_a \le 303$$

$$800 \le Z_b \le 1000 \tag{2.4}$$

The MO design optimization problem for the solar powered irrigation system is summarized as follows:

Max (f_1, f_2, f_3) subject to *design constraints* (2.5)

The Hyp-AP algorithm used in this work was programmed using the C++ programming language on a personal computer (PC) with an Intel dual core processor running at 2 GHz.

COMPUTATIONAL TECHNIQUE

Weighted Sum Analytical Programming (w-AP)

In MO optimization, meta-heuristic algorithms cannot be directly applied to a problem since most of these algorithms are developed for single-objective optimization. Many frameworks such as Normal-Boundary Intersection (NBI) (Das & Dennins, 1998) and the Fuzzy MO approaches (Pytel, 2012) are available and capable to be applied hand in hand with meta-heuristic algorithms. In this work, the AP method was integrated with the weighted sum approach (which is a function aggregation methodology with objective trade-offs). Algorithm 2.1 depicts the workflow of the w-AP approach:

Algorithm 2.1: Hypervolume-driven Analytical Programming (Hyp-AP)

Step 1: Initialize weights; w_1, w_2 and w_3 such that $(w_1 + w_2 + w_3) = 1$
Step 2: Initialize individual size N and population size, P.
Step 2: Randomly initialize the population vectors.
Step 3: Use instructions in population vectors to generate symbolic structures.
Step 4: Apply symbolic structures to generate solutions to the MO problem.
Step 5: Evaluate solutions against the aggregated function, F.
Step 7: If the $F^{new} = F^{old}$ or iter = T_{max} halt and print solutions, else proceed to step 8
Step 8: Mutate the vectors according to the degree of fitness, then insert mutated vector and go to Step 4.

Hypervolume-Driven Analytical Programming (Hyp-AP)

The AP algorithm is a generalized approach aimed towards the development of a universal method for symbolic regression. There are two known computing methods capable of symbolic regression - and hence able to synthesize tree-like symbolic structures (see Figure 1).

The first is known as genetic programming (GP) (Koza, 1990) and the second is grammatical evolution (GE) (Ryan *et al.*, 1998). The AP approach was first introduced in Zelinka, (2002) [7]. The core concept of this technique involves the generation of symbolic structures by evolutionary means (Zelinka *et al.*, 2005). Each instruction array organizes the instruction sequence in the symbolic structure (see Figure 2). On the other hand, each symbolic structure represents a computational expression (see Figure 3).

In this work, the hypervolume indicator (HVI) was embedded into the AP algorithm. The HVI measures the quality (degree of dominance) of the solutions in any given MO setting. Therefore the Hyp-AP utilizes the HVI to direct its execution towards higher dominance levels at successive iterations.

Figure 1. Tree-like symbolic structure

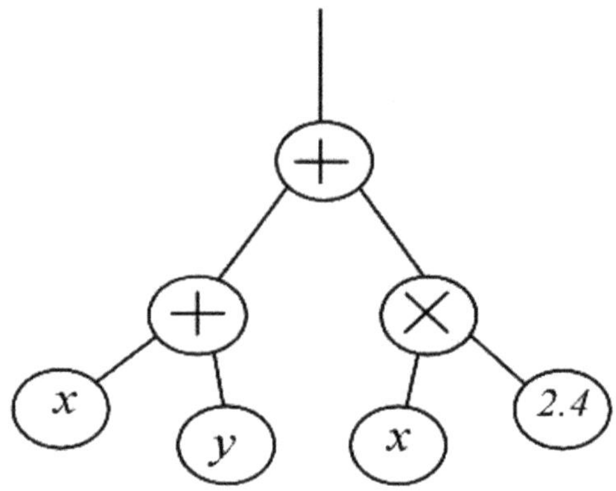

Figure 2. Instruction bit and the corresponding symbolic structure

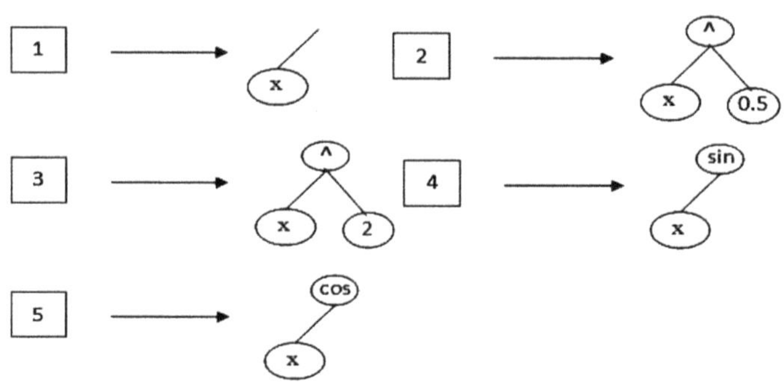

Figure 3. Symbolic structure representing a computational expression

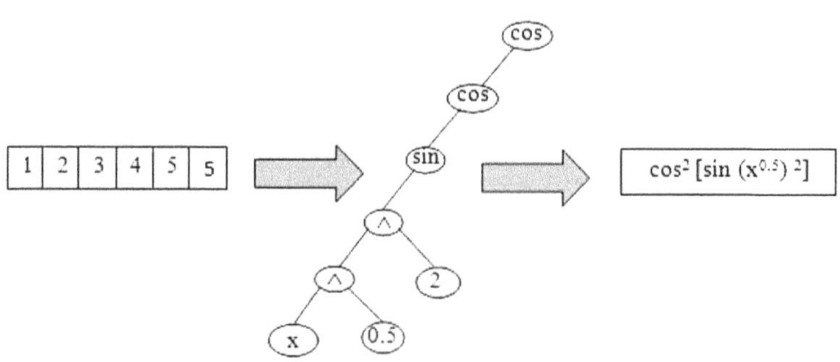

To drive the algorithm effectively towards dominance, the HVI was coupled with the mutation operator. This mechanism works such that, if at iteration, *i* the degree of dominance is lower than the previous iteration, *i-1*, then at iteration *i+1*, the mutation operator would increase its degree of mutation so that the successive population would have a more diverse solution pool. However, if the degree of dominance at *i* is higher than the previous iteration, *i-1*, then at iteration *i+1*, the mutation operator would decrease its degree of mutation so that algorithm may converge into the most dominant solution.

The original AP was modified by introducing an evolutionary mutation operator for this application. This mutation operator acts as a population diversifier during the iterations. AP starts by the initialization of a population of at least four individuals denoted as *P*. These individuals are real-coded vectors with s defined size, *N*. The initial population of individual vectors (the first iteration denoted $iter = 1$) are randomly generated for appropriate search ranges. These vectors contain instructions to generate the symbolic structures which in turn will become computational expressions. These symbolic structures will then be used to generate solutions to the MO optimization. The quality of these solutions (dominance) are then evaluated using the HVI. For the consequent iterations, the mutation operator is adjusted accordingly as described previously. Hence, the AP is driven by the HVI as it searches for highly dominant solutions in a self-driven way. Figure 4 and Algorithm 2.2 depict the work flow of the Hyp-AP approach:

Algorithm 2.2: Hypervolume-driven Analytical Programming (Hyp-AP)

Step 1: Initialize individual size *N, P*.
Step 2: Randomly initialize the population vectors.
Step 3: Use instructions in population vectors to generate symbolic structures.
Step 4: Apply symbolic structures to generate solutions to the MO problem.
Step 5: Evaluate solutions quality (dominance) using the HVI.
Step 7: If the fitness criterion is satisfied or $iter = T_{max}$, halt and print solutions else proceed to step 8
Step 8: Mutate the vectors according to the degree of fitness, then insert mutated vector into the initial population and go to Step 4.

A similar stopping criterion was set for both the w-AP and Hyp-AP algorithms. For the Hyp-AP, the variant $\varphi = \text{HVI}$ while for the w-AP, the variant $\varphi = F$.

Figure 4. Hyp-AP workflow

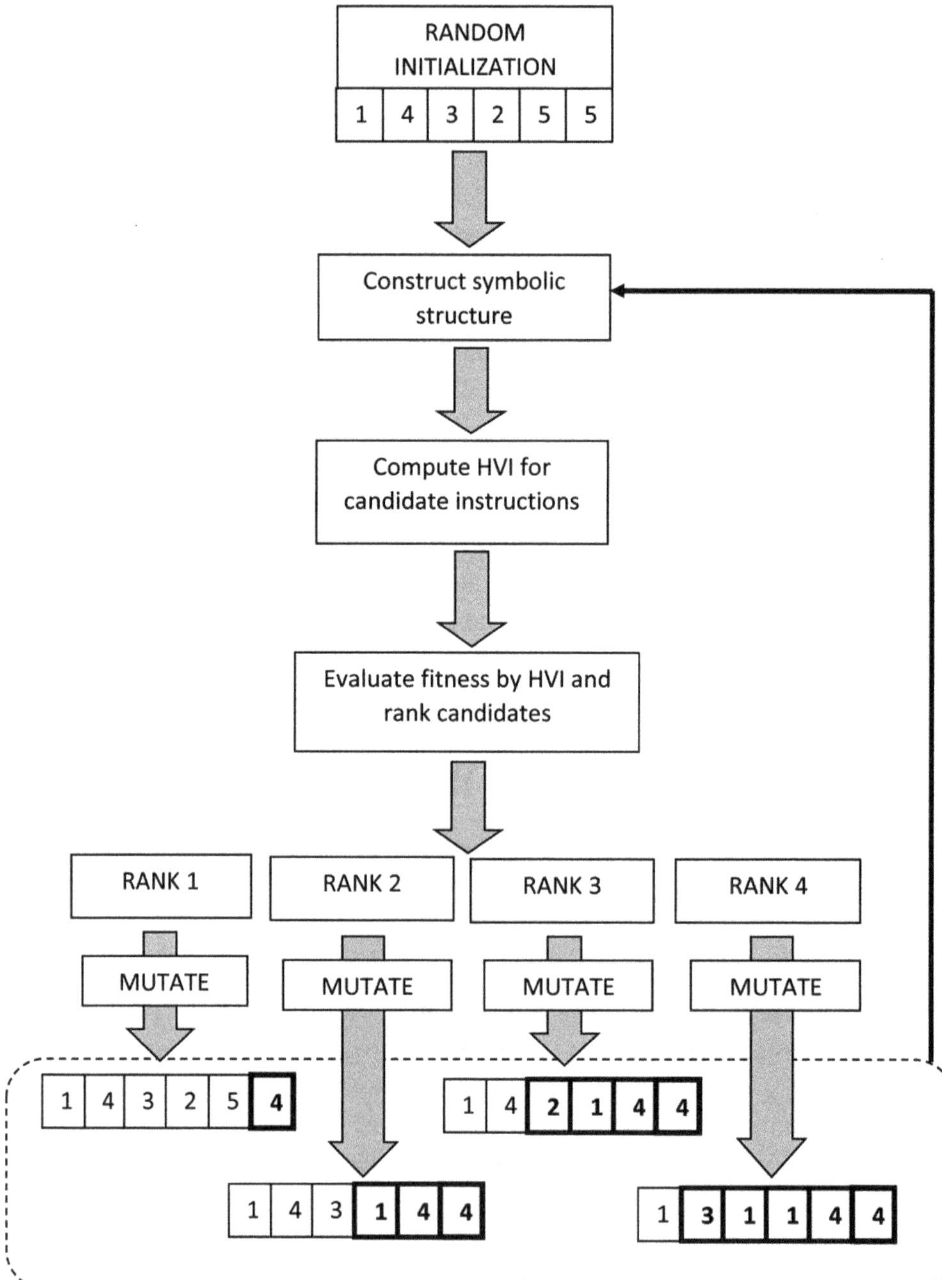

The following measure (equation 2.6) is used to compute the solution convergence and the stopping criterion is shown in Figure 5:

$$\varepsilon = \left(\frac{\varphi^{iter+1} - \varphi^{iter}}{\varphi^{iter}} \right) \times 100\% \tag{2.6}$$

The parameter setting for the w-AP and the Hyp-AP are given in Table 1.

NUMERICAL RESULTS

The solution sets (approximations of the Pareto frontier) were obtained using the w-AP and the Hyp-AP methods. The quality of these solutions was measured

Figure 5. Stopping criterion pseudo-code

```
IF (ε ≤ 5% or Tmax = iter) {
Halt and print solution}
ELSE {}
IF (ε ≥ 5% or Tmax > iter){
Mutate the vectors according to the degree of fitness
and re-insert mutated vector into the initial
```

Table 1. Parameter setting for the w-AP and the Hyp-AP algorithms

Parameters	Values
Initial population, *P*	4
Individual Size, *N*	6 bit
Mutation operator type	Bit-flip
Hypervolume reference point, *H*	(0, 0, 141253.8)

by using the value of the HVI. Since this problem is a maximization problem, hence the higher the HVI value the more dominant the solution. The values of the best, medium and worst individual solutions were identified using the HVI. These values are shown in Table 2.

The best, median and worst individual solutions were obtained at the weights (w_1, w_2, w_3) of (0.1, 0.1, 0.8), (0.5, 0.4, 0.1) and (0.3, 0.4, 0.3) with the computational times of 0.110, 0.103 and 0.078 seconds respectively. The value of the aggregated function for one of the weight sets, F with respect to the number of iterations in shown in Figure 6.

It can be seen from Figure 6 that the values of F increases in a almost regular step-wise fashion as the number of iterations increases. However, when the search reaches the near global optimum (at the 85th iteration), the solution converges at a maximum value and becomes stagnant with respect to the number of iterations. The best, medium and worst individual solutions of the Hyp-AP technique were identified using the HVI. These individual solutions along with their dominance values are shown in Table 4.

The best, median and worst individual solutions were obtained with the computational times of 0.212, 0.167 and 0.147 seconds respectively. Since the Hyp-AP is driven by the HVI, hence the HVI varies as the technique searches for more dominant (optimum) solutions in the objective space during the execution of the program. The value of the HVI for one of the runs with respect to the number of iterations is shown in Figure 7:

It can be seen from Figure 7 that the values of F increase irregularly as the number of iterations increase. However, when the search reaches the near

Table 2. The best, medium and worst individual solutions obtained by the w-AP method

	Best	Median	Worst
f_1	20.6838	20.6497	20.5553
f_2	16.8096	16.9326	16.7765
f_3	148103	147858	147181
x_a	0.721209	0.791193	0.894297
x_b	455.383	454.99	453.954
x_c	526.778	524.608	525.56
x_d	0.066075	0.059836	0.06103
Z_a	302.596	302.902	302.897
Z_b	812.385	809.592	810.663
HVI	2381390	2309195	2043987

global optimum (at the 109th iteration), the solution converges at a maximum value and becomes stagnant with respect to the number of iterations. In Chen *et al.* (1995), the Decision Support in the Design of Engineering Systems (DSIDES) software developed in (Mistree *et al.*, 1993) was implemented to the solar irrigation problem. The comparison of the individual best solutions obtained by the Hyp-AP and w-AP algorithms against the solutions obtained in Chen *et al.* (1995) is given in Table 4.

Figure 6. Value of the aggregated function, F with respect to the number of iterations

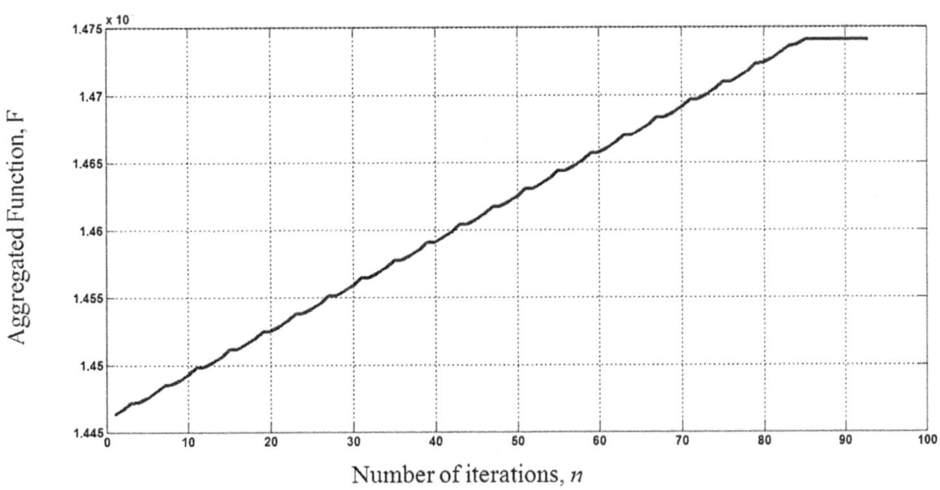

Table 3. The best, medium and worst individual solutions obtained by the Hyp-AP method

	Best	Median	Worst
f_1	20.7987	20.6639	20.7987
f_2	17.4495	16.7572	17.4495
f_3	148927	147960	148927
x_a	0.620651	0.761554	0.620651
x_b	456.5	455.184	456.5
x_c	524.661	525.57	524.661
x_d	0.038835	0.070007	0.038835
Z_a	302.707	302.799	302.707
Z_b	807.545	811.571	807.545
HVI	2784827	2322166	1902451

In Figures 6 and 7 the regularity of F and the irregularity of the HVI as the iterations increase were observed. This occurs since the HVI navigates the Hyp-AP program better through the objective space and attempts to successively obtain only highly dominant solutions. On the other hand, the w-AP only obtains solutions at compromised objectives (based on weights), hence is may not navigate the program as well as the Hyp-AP during its search in the objective space.

One of the advantages of the Hyp-AP method is that it obtains highly dominant solutions in MO scenarios. Besides, the Hyp-AP method also obtains optimal/dominant solutions that construct the Pareto frontier without compromising any of the objectives. For this solar irrigation application problem, the best individual solutions obtained by the Hyp-AP method were more dominant by the w-AP method and the DSIDES approach in Chen *et al.*, (1995) by 16.94% and 52.79% respectively.

Nevertheless, the Hyp-AP method has a drawback in terms of computational time as compared to the w-AP approach. This is because the Hyp-AP method contains the hypervolume segment which requires more computational effort as compared to w-AP. The w-AP method is more computationally efficient than the Hyp-AP method by approximately 48.42%.

In this work, both algorithms Hyp-AP and w-AP methods produced feasible solutions where no constraints of the model were broken. Besides, the Hyp-AP and w-AP methods performed stable calculations during the

Figure 7. Value of the HVI with respect to the number of iterations

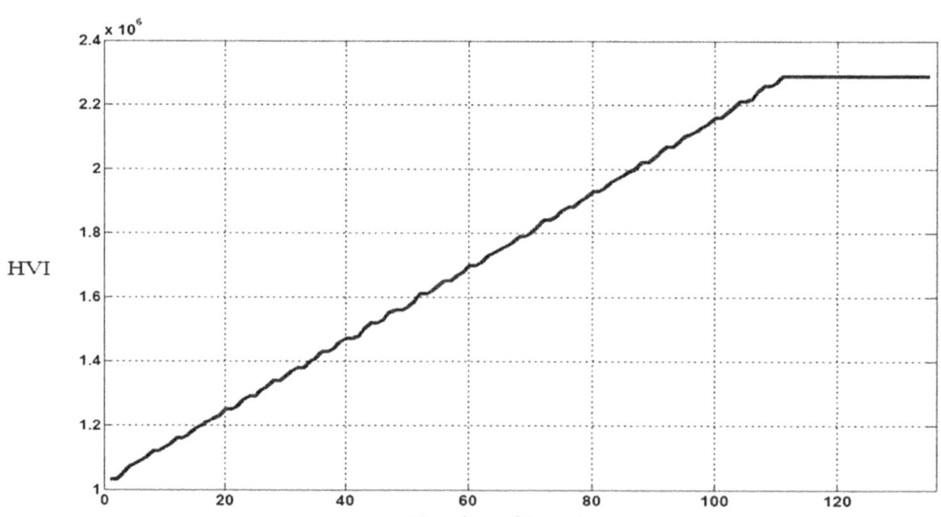

Table 4. The best individual solutions obtained by the w-AP, Hyp-AP and DSIDES method

	w-AP	Hyp-AP	DSIDES (Chen *et al.* (1995)
f_1	20.6838	20.7987	20.003
f_2	16.8096	17.4495	19.45
f_3	148103	148927	141143
x_a	0.721209	0.620651	3
x_b	455.383	456.5	450
x_c	526.778	524.661	550
x_d	0.066075	0.038835	0.0258
Z_a	302.596	302.707	0.02577
Z_b	812.385	807.545	0.02577
HVI	2381390	2784827	1822616

program executions. A new local optimum was discovered in this work by the implementation of the Hyp-AP method (see Tables 3 and 4).

PART 2: DESIGN OF SOLAR POWERED IRRIGATION SYSTEM IN CONSIDERATION OF TRANSIENT WEATHER CONDITIONS

Background

Diesel generators, gas turbines and other fossil-fuel based power systems have been widely employed for powering conventional irrigation systems. Currently, various issues related to the utilization of fossil fuel in power systems have surfaced (e.g. price fluctuations, environmental concerns and efficiency). Hence in recent studies, the harnessing of solar energy to power irrigation pumps have become popular (Helikson *et al.*, 1991; Wong & Sumathy, 2001; Jasim *et al.*, 2014; Dhimmar *et al.*, 2014). The design and sizing of solar power systems greatly impacts the system's reliability, emissions and efficiency (Al-Ali *et al.*, 2001). Thus challenges in effective sizing of this system falls into the realm of optimization. Here the optimal sizing and design are obtained such that the following targets are achieved (Shivrath *et al.*, 2012; Carroqino *et al.*,2009):

- Increased savings
- Low emissions
- Good system efficiency
- High power output

Efforts in optimizing solar powered irrigation systems have been performed via the implementation of metaheuristics such as: genetic algorithms (GA) and particle swarm optimization (PSO) (Gouws & Lukhwareni, 2012). However, due to the complexity of the system, the design needs to be carried out in such a way that it takes into account multiple aims simultaneously (i.e. multiobjective optimization) (Chen *et al.*, 1995).

Real-world optimization problems often contain large degrees of uncertainties. To effectively handle such problems, higher order fuzzy logic (FL) such type-2 FL approaches are often employed in tandem with optimization techniques (Castillo and Melin, 2012; Ontoseno *et al.*, 2013; Sánchez *et al.*, 2015). Most existing research works involving the application of type-2 FL systems revolve around control theory and control system design (Fayek *et al.*, 2014; Martinez *et al.*, 2011; Bahraminejad *et al.*, 2014; Oh *et al.*, 2011; Linda and Manic, 2011). For instance in Wu and Tan (2004), the authors investigated the effectiveness of evolutionary type-2 FL controllers for uncertainty modelling in liquid-level processes. In that work, the authors employed the genetic algorithm (GA) to evolve the type-2 FL controller. This approach was found to perform very well for modeling uncertainties in complex plants compared to conventional type-1 FL frameworks. In Bahraminejad *et al.*, (2014), a type-2 FL controller was employed for pitch control in wind turbines. Pitch control in wind turbines are critical for power regulation and reduction of fatigue load in the components of the turbine. In Bahraminejad *et al.*, (2014), the type-2 FL controller was shown to significantly improve the adjustment of the pitch angle, rotor speed and power output of the wind turbine generator. Similarly, in Allawi (2014), a type-2 FL controller was utilized for controlling robots involved in cooperation and target-achieving tasks in multi-robot navigation systems. In that work, the controllers were optimized using the Particle Swarm Optimization (PSO) and the Hybrid Reciprocal Velocity Obstacles techniques. The author discovered that the optimized type-2 FL controller performed very well for controlling such robots.

Besides control theory and engineering, type-2 FL has also been employed for modelling systems endowed with high levels of uncertainty. For instance in Paricheh and Zare (2013), a type-2 FL system was used to predict long-term traffic flow volume. In Paricheh and Zare (2013), the traffic flow data was heavily influenced by various time-dependent uncertainties and nonlinearities. In that work, the authors employed a type-2 FL system in combination with genetic algorithm and neural net-based approaches. Another interesting implementation of type-2 FL was presented in the work of Qiu *et al.*, (2013). In that research, the authors focused on developing a general interval type-2 fuzzy C-means algorithm. The proposed fuzzy algorithm was employed for medical image segmentation in magnetic-resonance images (MRI). These MRIs are usually noisy and highly inhomogeneous. The works of Castillo and Melin (2012) and Dereli *et al.*, (2011) provides a more comprehensive review on type-2 FL systems applied in industrial settings.

Swarm intelligence (SI) stands as one of the most favored strategies for solving complex optimization problems due to its effectiveness during search operations and in terms of computational cost (Liu *et al.*, 2002). Some of the most popular SI-based techniques are, cuckoo search (CS) (Yildiz, 2013), ant colony optimization (ACO) (El-Wahed, 2008), PSO (Kennedy and Eberhart, 1995) and bacterial foraging algorithm (BFA) (Passino, 2002). In many past research works, PSO has been implemented extensively for solving nonlinear optimization problems. Recently, alternative optimization strategies such as BFA has been implemented for such purposes. BFA's computational performance has been demonstrated to be as good as and sometimes better than other SI-based techniques (Al-Hadi and Hashim, 2011). BFA is inspired by the natural behavior of the E. Coli bacterium. This behavior involves the search mechanisms utilized by the bacteria during nutrient foraging. These mechanism were then employed by Passino (2002) to design the BFA for solving complex optimization problems. The central principle of the BFA framework is as follows:

Each bacteria in the swarm tries to maximize its energy per unit time spent during the foraging for nutrients while simultaneously evading noxious substances

In recent times, BFA has been seen applied in many engineering applications (e.g. economic dispatch, engineering design, manufacturing technology, power systems and control systems). In the research work by Mezura-Montes and Hernandez-Ocana (2009), specific adjustments to the BFA approach was performed to enhance its optimization capability. This enhanced-BFA was then effectively implemented for engineering design. In Mezura-Montes *et al.*, (2014), the design optimization of a crank-rocker-slider (variable transmission) system was performed using BFA.

The design optimization problem in Mezura-Montes *et al.*, (2014) was formulated in two forms (single-objective and bi-objective). The authors then restructured the conventional BFA for multiobjective optimization. The conventional and multiobjective techniques were successfully implemented to optimize the mechanical design of the crank-rocker-slider system. BFA has also been applied in cellular manufacturing systems (Nouri and Hong, 2013). In these systems, the cell formation issue is tackled while considering the number of exceptional elements and cell load variations. In Nouri and Hong (2013), using the BFA approach, part families and machine cells were generated by the authors. In Panda *et al.*, (2009), the BFA was seen to aid the manufacturing process of rapid prototyping. In that work, the BFA was used for optimizing the process parameters employed for fused deposition modeling (FDM).

Besides manufacturing, BFA has also been widely applied by engineers/researchers in power engineering and distribution. In such applications the BFA is utilized for optimizing economic load dispatch. In economic load dispatch, the target is to obtain the most optimal load dispatch for the power generating units while taking into account variable load demands and load constraints (Vijay, 2012). In addition, BFA has been utilized for obtaining optimal power flow (i.e. economic and efficient) in flexible alternating current transmission system (FACTS) devices (Ravi *et al.*, 2014).

OVERVIEW OF TYPE-2 FUZZY LOGIC

Type-2 fuzzy sets are generalizations of the conventional or type-1 fuzzy sets (Zadeh, 1975). The primary feature of the type-1 fuzzy set is its membership function, $\eta_F(x) \in [0,1]$ and $x \in X$. Type-2 FL employs a membership function of a second order, $\mu_F(y, \eta_F(x)) \in [0,1]$ such that $y \in Y$. Therefore, $\mu_F(y, \eta_F(x))$ is a membership function that requires three-dimensional inputs. The type-2 fuzzy set is defined as follows:

$$\tilde{F} = \left\{(y, \eta_F(x)), \mu_F(y, \eta_F(x)) : \forall x \in X, \forall y \in Y, \eta_F(x) \in [0,1]\right\} \tag{2.7}$$

The type-2 membership function has two membership grades: primary and secondary memberships. Thus, a crisp set (or function) undergoes fuzzification twice such that the first fuzzification transforms it to a type-1 fuzzy set (via the primary membership function). Using the secondary membership function the type-1 fuzzy set is transformed to a type-2 fuzzy set. In other words: the type-2 fuzzy set results from the fuzzification of a type-1 fuzzy set. This operation aims to improve its efficacy and accuracy in capturing uncertainties. The region covered by the type-1 fuzzy sets in type-2 FL systems is represented by the footprint of uncertainty (FOU). This region of uncertainty is contained by the uppermost and lowermost type-1 membership functions $\eta_F^U(x)$ and $\eta_F^L(x)$ respectively. A type-2 FL system usually consists of four subcomponents: fuzzifier, inference engine, type reducer and defuzzifier. The fuzzifier directly transforms the crisp set into a type-2 fuzzy set. The inference engine functions to combine rules to map the type-2 fuzzy set from crisp inputs. Therefore each rule is interpreted as a type-2 fuzzy implication in the inference engine. In this work, all the consequent and antecedent sets are generalized type-2 fuzzy sets. The rule, R_i from a type-2 FL system could be generally represented as follows:

R_i: IF x_1 is $\tilde{M}_1$ AND....AND x_j is $\tilde{M}_j$ THEN y_1 is $\tilde{N}_1$,..., y_k is $\tilde{N}_k$ such that $i \in [1, Z]$ (2.8)

where j is the number of fuzzy inputs, k is the number of fuzzy outputs and i is the number of rules. The type reducer functions to transform (or reduce) the type-2 fuzzy set to a type-1 fuzzy set. Various type-reduction approaches have been developed in the past. For instance: centroid type reduction (Mendel and John, 2002) vertical slice-centroid type reduction (Lucas *et al.*, 2007), alpha cuts/planes (Hamrawi and Coupland, 2009) and the random sampling technique (Greenfield *et al.*, 2005). Defuzzification on the other hand reduces the type-1 fuzzy set to a crisp output similar to operations in conventional type-1 FL systems. There are various defuzzification techniques which are employed selectively to suit specific data representations and applications (Rao and Saraf, 1996).

TYPE-2 FUZZY LOGIC FOR NOISE MODELLING

Energy systems that rely on their surroundings are often difficult to design if the surroundings are noisy. In solar-powered systems, weather-dependent variables such as insolation and ambient temperature are often found to be irregular and noisy. Thus, when designing such systems, the model utilized should be able to account for such irregularities. In this work, equipped with meteorological data, type-2 FL is employed to model and incorporate insolation and ambient temperature into the optimization formulation. The meteorological data for Santa Rosa Station at California was retrieved from the weather database of the University of California Agriculture and Natural Resources. The daily average insolation (in W/m^2) and ambient temperature (in K) was obtained for every month of the year 2014. Table 5 provides the monthly ambient temperature and insolation data:

The primary membership function, $\eta_F(x)$ was employed to model the monthly data while the secondary membership function, $\mu_F(y, \eta_F(x))$ was used to model the overall annual data. This way the noisy monthly fluctuations in the data is taken into consideration via the type-2 FL approach. The overall type-2 fuzzy modelling strategy is given in Figure 8:

The S-curve function is employed as the primary and secondary membership functions ($\eta_F(x)$ and $\mu_F(y, \eta_F(x))$). Therefore type-2 fuzzification is performed on the ambient temperature (Z_a) and insolation, (Z_b) using the S-curve membership function. This is carried out by determining the average, maximum and minimum values of insolation and ambient temperature from the meteorological data. The S-curve membership function is as follows:

Table 5. Monthly average ambient temperature and insolation taken at Santa Rosa Station, California in 2014

Month, *m*	Ambient Temperature, Z_a (K)			Average Solar Radiation, Z_b (W/m^2)		
	Max	Min	Average	Max	Min	Average
1 (Jan)	297.4	265.2	281.3	146	43	104.71
2 (Feb)	295.8	266.3	281.05	186	16	107.11
3 (Mac)	301.9	272.4	287.15	236	41	169.16
4 (April)	304.7	273.6	289.15	295	77	240.03
5 (May)	305.8	275.2	290.5	335	30	286.52
6 (Jun)	306.3	276.3	291.3	336	211	306.20
7 (July)	308	279.1	293.55	330	102	255.16
8 (Aug)	306.9	279.7	293.3	282	62	220.71
9 (Sept)	308	278.6	293.3	252	92	196.13
10 (Oct)	309.1	273.6	291.35	216	27	149.87
11 (Nov)	301.3	271.9	286.6	146	18	99.67
12 (Dec)	293.6	270.8	282.2	115	14	62.45

Figure 8. Type2 fuzzy modelling strategy

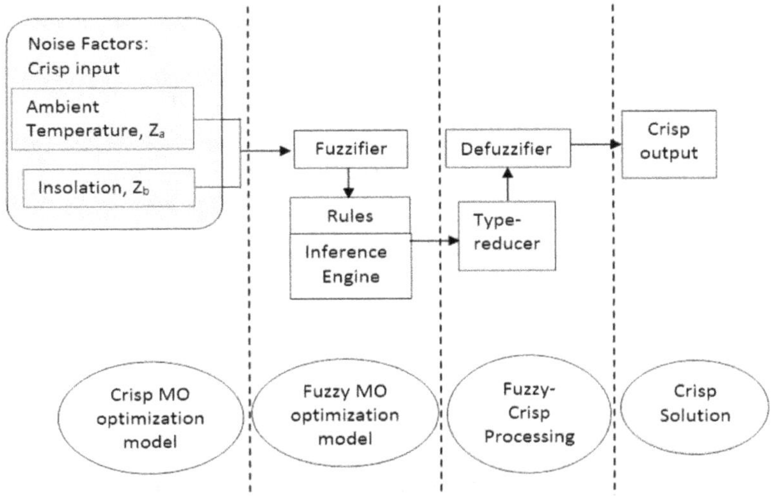

$$\mu_{\bar{b}_i} = \begin{cases} 1 & if\ b_i \leq b_i^a \\ \dfrac{B}{1 + Ce^{\alpha\left(\frac{b_i - b_i^a}{b_i^b - b_i^a}\right)}} & if\ b_i^a \leq b_i \leq b_i^b \\ 0 & if\ b_i \geq b_i^b \end{cases} \tag{2.9}$$

where B and C are parameters which are tuned heuristically such that the membership fits the meteorological data effectively. Using Zadeh's extension principle, all crisp variables (ambient temperature (Z_a) and insolation (Z_b)) and their respective constraints are transformed via type-2 fuzzification. Assuming a credibility level ε, ($0 < \varepsilon < \frac{B}{1+C}$) chosen by the Decision Maker (DM), as he/she takes a risk and ignores all the membership degrees smaller than the ε levels (Rommelfanger, 1989). The FOU is the union of all the primary memberships (Mo *et al.*, 2014). In this case, the union of all the primary S-curve memberships, $\eta_F(x)$ for each month depicts the FOU:

Let $\eta_F^i(x) \in \left(J_x^i \subseteq [0,1]\right)$ such that $i = [L,U]$, THEN

$$FOU = \bigcup_{x \in X} J_x^i \tag{2.10}$$

where J_x^i is the fuzzy set, L is the lower bound and U is the upper bound. A graphical depiction of the FOU generated by the primary S-curve memberships in this work is given in Figure 9.

There are many readily available techniques for type-reduction and defuzzification. In this work, the alpha-plane approach (Hamrawi and Coupland, 2009; Liu 2006) was employed for type-reduction while the conventional alpha-cut approach (Klir and Yuan, 1995) was used for the defuzzification. An alpha cut can be defined on a fuzzy set, $\tilde{F}$ via its decomposed form as follows:

$$\tilde{F} = \bigcup_{\alpha \in [0,1]} \alpha \cdot F_{\alpha} \tag{2.11}$$

where F_{α} is an $\alpha-$level set. Similarly, an alpha-cut on a type-2 fuzzy set could be performed via the decomposition theorem. Since this operation is performed on a type-2 fuzzy set, it is defined as an alpha-plane instead of an alpha cut:

$$\tilde{\tilde{F}} = \bigcup_{\tilde{\alpha} \in [0,1]} \tilde{\alpha} \cdot \tilde{F}_{\tilde{\alpha}} \tag{2.12}$$

where $\tilde{F}_{\tilde{\alpha}}$ is a type-2 $\alpha-$level set. It should be noted that by using Zadeh's extension principle, the alpha-planes could be utilized to execute type-2 fuzzy

Figure 9. FOU generated by the primary S-curve memberships

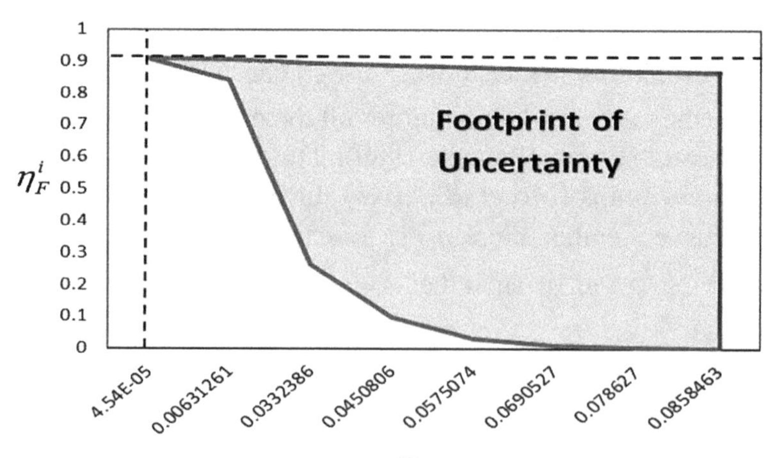

operations using interval type-2 fuzzy sets. This is analogous to implementations in type-1 fuzzy sets since the extension principle could be evoked to extend functions that interrelate crisp, type-1 fuzzy as well as type-2 fuzzy sets.

SOLAR-POWERED IRRIGATION SYSTEM

The design of the solar-powered irrigation system significantly affects its system characteristics (see Part 1). Besides design parameters, noisy environmental factors such as the ambient temperature, Z_a (K) and the level of insolation, Z_b (W/m^2) greatly influences the system's characteristics. The objective functions and constraints are thus formulated such that they consider the design variable while taking into account the environmental factors. The design formulation is given as given in equations 2.1 - 2.4:

As mentioned in previous sections, to accurately account for the uncertainties arising from the noisy environment, the fuzzy type-2 approach is incorporated into the system design. Taking this view, the ambient temperature, Z_a (K) and the level of insolation, Z_b (W/m^2) is fuzzified. By implication, the target objectives and the constraints that bound the environmental factors (Z_a and Z_b) in the model above is transformed to a fuzzy form. The type-1 fuzzy constraints for the environmental factors are as follows:

$$\tilde{Z}_a^m \in [Z_{a,\min}^m, Z_{a,\max}^m]$$

$$\tilde{Z}_b^m \in [Z_{b,\min}^m, Z_{b,\max}^m] \tag{2.13}$$

where m represents the months in the year as in Table 5 while $Z_{a,\max}^m$ and $Z_{b,\max}^m$ are the maximum monthly average and $Z_{b,\max}^m$ and $Z_{b,\max}^m$ are the minimum monthly average. The type-2 fuzzy constraint for the whole year is represented as follows:

$$\tilde{\tilde{Z}}_a \in [\tilde{Z}_{a,\min}, \tilde{Z}_{a,\max}] = [265.2, 309.1]$$

$$\tilde{\tilde{Z}}_b \in [\tilde{Z}_{b,\min}, \tilde{Z}_{b,\max}] = [14, 336] \tag{2.14}$$

Therefore, the MO design optimization of the solar-powered irrigation system is effectively transformed to a fuzzy MO optimization problem:

$$\begin{gathered}Maximize \rightarrow (\tilde{\tilde{f}}_1, \tilde{\tilde{f}}_2, \tilde{\tilde{f}}_3)\\ subject\ to\ Crisp\ Design\ Constraints\ \&\ Type-2\ Fuzzy\ Environmental\ Constraints\end{gathered} \tag{2.15}$$

BACTERIA FORAGING ALGORITHM

The dynamics of bacteria foraging is directly influenced by evolutionary biology. Thus, bacteria with successful foraging strategies would stand a better chance in propagating their genetic makeup as compared to bacteria with poor strategies. This way bacteria at successive generations always contain improved foraging strategies relative to past generations and the strategies continually improves as they go along reproducing. Due to such progressive behavior, many researches were targeted to model bacteria foraging dynamics as an optimization process. The central theme of foraging viewed from this perspective is that the organisms conduct the search in such a way that they maximize the energy they obtain from the nutrients at minimal time during foraging.

Foraging efforts vary according to the species of the organism and the environment where the foraging is taking place. For instance, herbivores would find it easier to locate food as compared to carnivores in any habitat. As for the environmental factor, the distribution of nutrients in desert or tundra conditions are sparser in contrast with the nutrient-rich tropical jungles. Design of effective and efficient search strategies for nutrient foraging which respects the previous constraints is critical for the long-term well-being of any organism. Another important factor to be considered for the design of effective search strategies is the type of nutrient. The type of nutrient will influence the fractionalization and planning of the strategy (O'Brien *et al.*, 1990). For instance, consider a case where the nutrient is stationary but hidden in a hard shell (e.g. eggs). Then the organism would have to design the foraging strategy in such a way that it searches for the shell (1), evades the nutrients parent(s) (2), breaks the shell (3), consumes the nutrient (4) then escapes the nutrient location or nest before it gets attacked/killed (5).

In many organisms, synergetic foraging strategies are observed to emerge in nature (e.g. ants, bees and thermites). These organisms create communication mechanisms that enable them to share information about the foraging efforts led by each individual in the group. Such mechanisms provide the capability of the organisms to conduct 'group/swarm foraging'. Group foraging provides these organisms with a plethora of advantages such as increased protection against predators and enhanced hunting/foraging strategies. These advantageous traits increases the organism's chances for finding nutrients in good time. Besides synergetic strategies for foraging, other strategies such as cooperative building (Turner, 2011), group defense (Schneider and McNally, 1992) and other cooperative group behaviors are common in nature.

In the BFA, four main levels of loops are present in the technique (chemotaxis, swarming, reproduction and elimination-dispersal loops). These loops manage the main functional capabilities of the BFA. Each of the mentioned loops are designed according to bacteria foraging strategies and principles from evolutionary biology. These loops are executed iteratively until the total number of iterations, N_T is satisfied. Each of the main loops may be iterated until some fitness condition is satisfied or until a user-defined loop cycle limit (chemotaxis (N_c), swarming (N_s), reproduction (N_r) and elimination-dispersal (N_{ed})) is reached. In chemotaxis, the bacteria with the use of its flagellum, swims and tumbles towards the nutrient source. The tumbling mode allows bacterium motion in a fixed direction while the tumbling mode enables the bacterium to augment its search direction accordingly. Applied simultaneously, these two modes give the bacterium capability to stochastically move towards a sufficient source of nutrient. Thus, computationally chemotaxis is presented as follows:

$$\theta^{i}(j+1,k,l,m) = \theta^{i}(j,k,l,m) + C(i)\frac{\Delta(i)}{\sqrt{\Delta(i)\Delta^{T}(i)}} \tag{2.16}$$

where $\theta^{j}(j+1,k,l,m)$ is the ith bacterium at the jth chemotactic step, k^{th} swarming step and l^{th} reproductive step and m^{th} elimination-dispersal step. *C(i)* is the size of the step taken in a random direction which is fixed by the tumble, and $\Delta \in [-1,1]$ is the random vector.

In the swarming phase, the bacterium communicates to the entire swarm regarding the nutrient profile it mapped during its movement. The communication method adopted by the bacterium is cell-to-cell signaling. In *E.Coli* bacteria, aspartate is released by the cells if it is exposed to high amounts of succinate. This causes the bacteria to conglomerate into groups and hence move in a swarm of high bacterial density. The swarming phase is mathematically presented as follows:

$$J(\theta,P(j,k,l,m)) = \sum_{i=1}^{S}\left[-D_{att}\exp(-W_{att}\sum_{m=1}^{P}\left(\theta_m - \theta_m^i\right)^2\right] + \sum_{i=1}^{S}\left[-H_{rep}\exp(-W_{rep}\sum_{m=1}^{P}\left(\theta_m - \theta_m^i\right)^2\right] \tag{2.17}$$

where $J(\theta,P(j,k,l,m))$ is the computed dynamic objective function value (not the real objective function in the problem), S is the total number of bacteria, P is the number of variables to be optimized (embedded in each bacterium), while H_{rep}, W_{rep}, H_{att}, and W_{att} are user-defined parameters.

During reproduction, healthy bacteria or those which are successful in securing a high quantities of nutrients are let to reproduce asexually by splitting into two. Bacteria which do not manage to perform according to the specified criteria are eliminated from the group and thus not allowed to reproduce causing their genetic propagation (in this case their foraging strategies) to come to a halt. Due to this cycle, the amount of individual bacterium in the swarm remains constant throughout the execution of the BFA. Catastrophic events in an environment (such as a sudden change in physical/chemical properties or rapid decrease in nutrient content) can result in the annihilation of a population of bacteria. Such events can cause bacteria to be killed and some to be randomly dispersed to different locations in the objective space. These events which are set to occur in the elimination/dispersal phase help to maintain swarm diversity to make sure the search operation is efficient. Figure 10 shows the work flow of the BFA technique. The pseudo-code for the BFA approach is provided below (Algorithm 2.3):

Algorithm 2.3: Bacteria Foraging Algorithm (BFA)

```
START PROGRAM
Initialize all input parameters (S, P, H_rep, W_rep, H_att, W_att, N_T,
N_c, N_r, N_s, N_ed)
Generate a randomly located swarm of bacteria throughout the
objective space
Evaluate bacteria fitness in the objective space
```

Figure 10. The work flow of the BFA

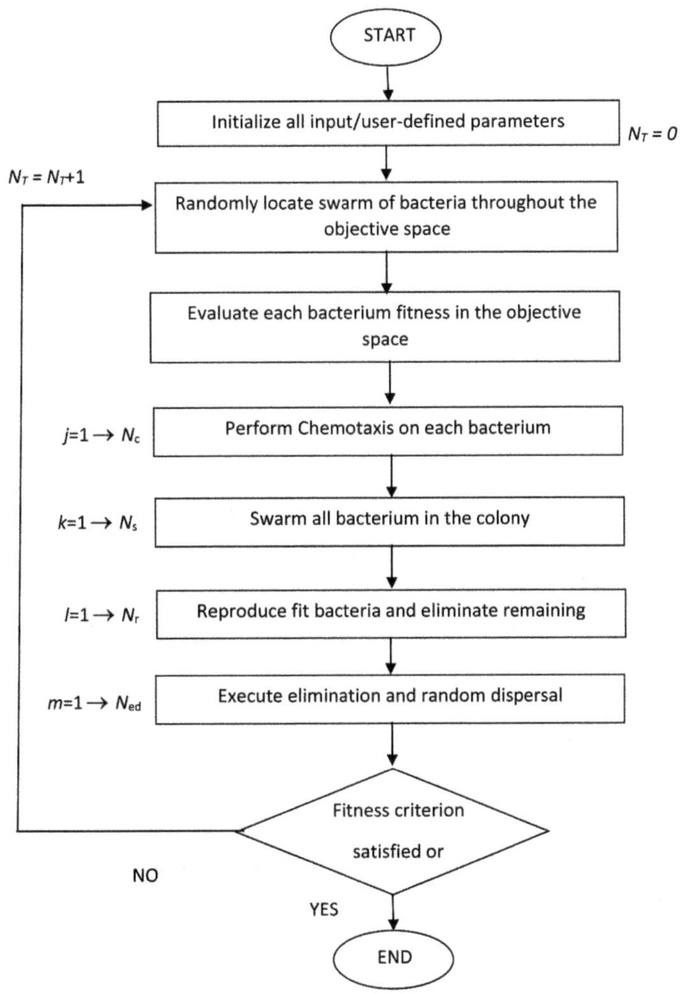


```
For i=1 → N_T do
For l=1 → N_r do
For m=1 → N_ed do
For j=1 → N_c do
For k=1 → N_s do
Perform chemotaxis - bacterium swim and tumble until maximum
fitness/loop cycle limit is reached
Perform swarming - bacterium swarm until maximum fitness/loop
cycle limit is reached
End For
                                   End For
                 If bacterium healthy/maximally fit then split
```

```
and reproduce
Else eliminate remaining bacterium
                End For
Execute catastrophic elimination by assigning some probability
of elimination to the swarm. Similarly disperse the remaining
swarm randomly.
End For
End For
END PROGRAM
```

SIGMA DIVERSITY METRIC

The diversity measure used in this work is the sigma diversity metric (Mostaghim and Teich, 2003). The Sigma Diversity Metric (SDM) evaluates the locations of the solution vectors in the objective space relative to the sigma vectors. For lower dimensional objective spaces ($n < 3$), metrics that are based on spherical and polar coordinates could be used. However, as the dimensions increase beyond three ($n \geq 3$), the mentioned coordinate systems do not define the distribution of the solution vectors well (Mostaghim and Teich, 2005). In such scenarios, the SDM is highly effective for computing the solution distribution. To begin the computation of the SDM, two types of sigma lines would have to be constructed. First the sigma lines that represent the solution vectors, σ' and the sigma lines that represent the reference lines, σ. The sigma lines that represent the solution vectors can be computed as the following:

$$\sigma'_k(ij) = \frac{f_i^2 - f_j^2}{\sum_{l=1}^{n} f_l^2} \text{ such that } \forall i \neq j \tag{2.18}$$

where k denotes the index that represents the number of solution vectors, i, j and l denotes the index that represents the number of objectives and n denotes the total number of objectives. Then the magnitude sigma $|\sigma'_k|$ is computed as follows:

$$|\sigma'_k| = \sqrt{\sum_{i=1}^{m}\sum_{j=1}^{m} \sigma'_k(ij)} \tag{2.19}$$

Thus, for each line in the objective space (solution vector or reference line), there exists a unique sigma value. The central working principle is that the inverse mean distance of the solution vectors from the reference sigma vectors are computed. Since the reference sigma vectors are distributed evenly along the objective space, the inverse mean distance depicts the diversity of the solution spread. High values of the sigma diversity metric, indicates high uniformity and diversity in terms of the distribution of the solution vectors in the objective space.

COMPUTATIONAL OUTCOME AND ANALYSIS

In Part 2, all computational procedures (algorithms and metrics) were developed using the Visual C++ Programming Language on a PC with an Intel i5-3470 (3.2 GHz) Processor. The compromised solutions were obtained using the BFA and utilized for the construction of the Pareto frontier. The fuzzy MO design problem was converted to a scalarized aggregate single-objective form using the weighted sum framework. Hence for various scalarization, the compromised solutions to the MO problem were obtained. In this work, each Pareto frontier was constructed using a cumulative of 35 solution points. The fuzzified ambient temperature, Z_a and insolation, Z_b is shown in Figures 11 and 12 respectively:

As in equations 2.9 and 2.10, the $\mu_{F,S}$ and $\mu_{F,T}$ are the primary membership grades while $\eta_{F,S}$ and $\eta_{F,T}$ are the secondary membership grades. The S and T subscripts denote the insolation and ambient temperature respectively. When performing type-2 fuzzy modeling, the number of membership grades are often large. Thus, data acquisition may become rather complicated as compared to data obtained using a fuzzy type-1 systems. To reduce the complexity of data acquisition, some approximate symmetry properties of the fuzzy membership grades are exploited. Thus, data acquisition could be performed by fixing the insolation membership grades and letting the ambient temperature membership grades vary. The other scenario being fixing the ambient temperature membership grade and varying the insolation grades. Due to the symmetrical properties of the membership grades, the final results should mirror each other in both scenarios. Evoking the symmetry property, the mapping of the objectives to the membership grades are represented in terms of membership grades of the ambient temperature while the membership grades of the insolation are let to vary in the following ranges:

Figure 11. Fuzzy ambient temperature (Z_a) versus membership grades

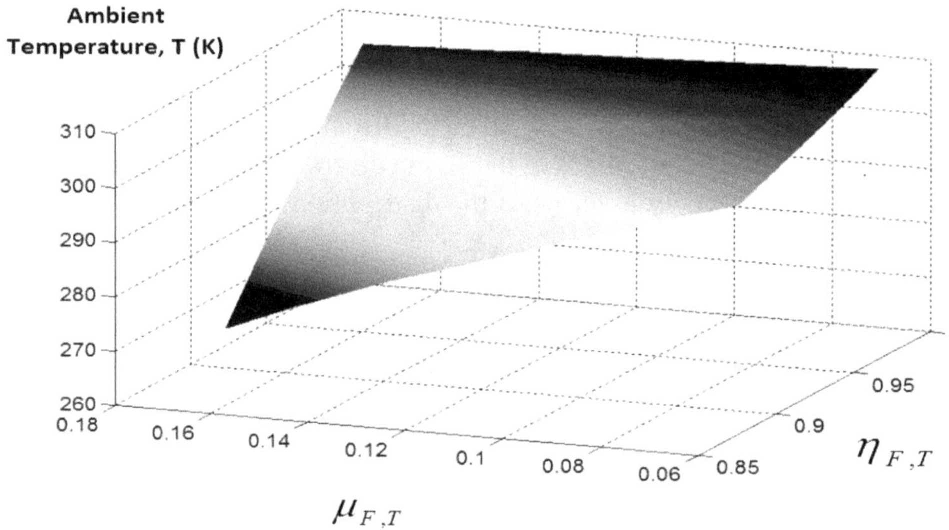

Figure 12. Fuzzy insolation (Z_b) versus membership grades

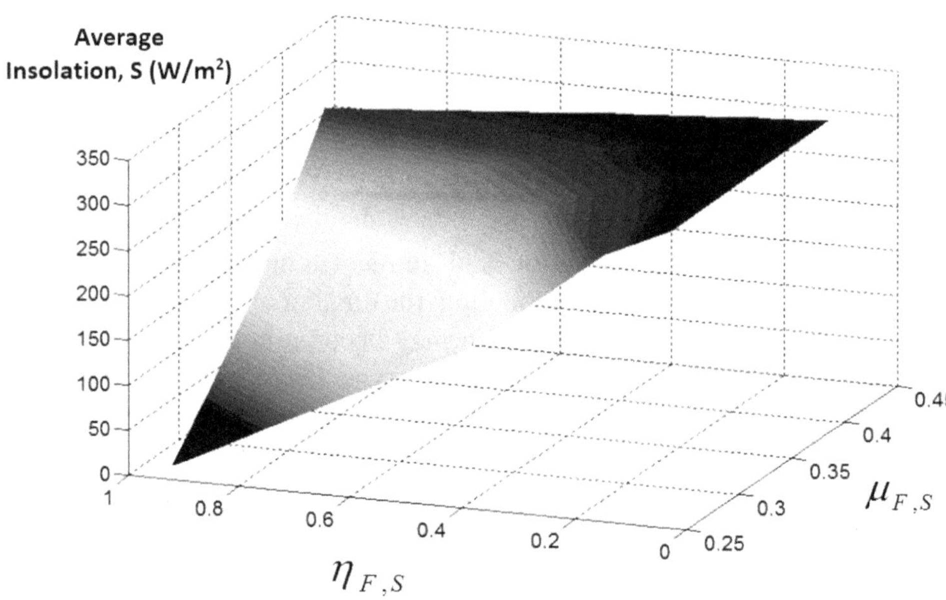

$\mu_{F,S} = [0.25264\ ,0.39913]$

$$\eta_{F,S} = [0.02907\ ,0.92274] \tag{2.20}$$

The mapping of the power output, f_1 (kW) to various primary and secondary membership grades (of ambient temperature) are depicted in Figure 13:

Referring to Figure 13, the maximal variation of power outputs at all membership grades is 0.3671 kW. A maximum power output of 20.1267 kW was obtained at $\mu_{F,T} = 0.9871$ and $\eta_{F,T} = 0.0665$. The minimal power output at $\mu_{F,T} = 0.8967$ and $\eta_{F,T} = 0.1717$ was 19.7596 kW. Figure 2.14 and 2.15 show the mapping of the overall efficiency, f_2 (%) and the fiscal savings, f_3 (USD) respectively relative to the primary and secondary membership grades (ambient temperature):

In Figure 14, the maximal overall efficiency of 17.9509% was obtained at $\mu_{F,T} = 0.9871$ and $\eta_{F,T} = 0.0665$ while the minimal overall efficiency of 16.7487% was obtained at $\mu_{F,T} = 0.8967$ and $\eta_{F,T} = 0.1717$. The highest variation in overall efficiency is 1.022%. Referring to Figure 15, the fiscal

Figure 13. Power output, f_1(kW) relative to membership grades

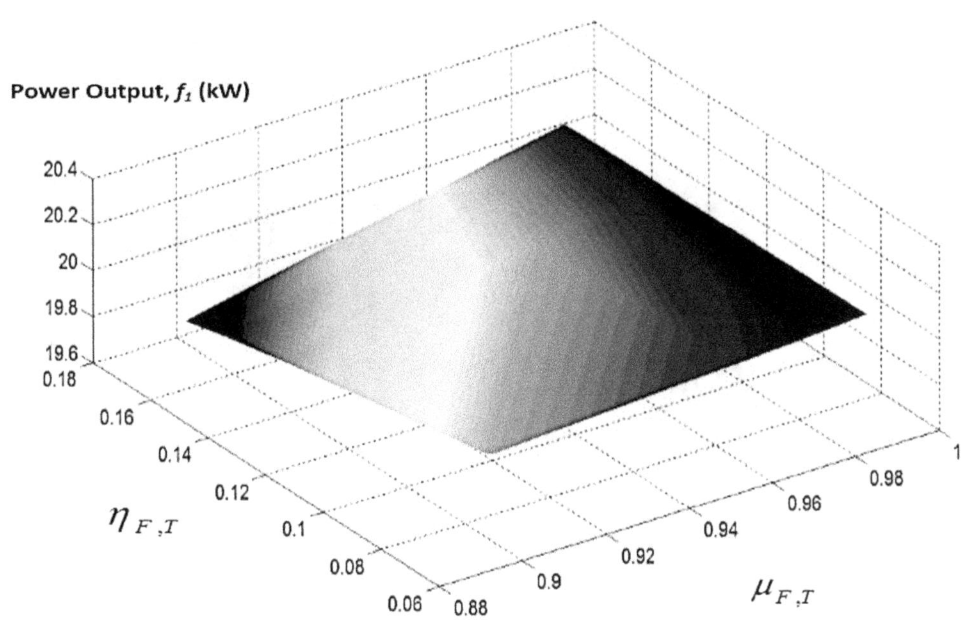

Figure 14. Overall efficiency, f_2(%) relative to membership grades

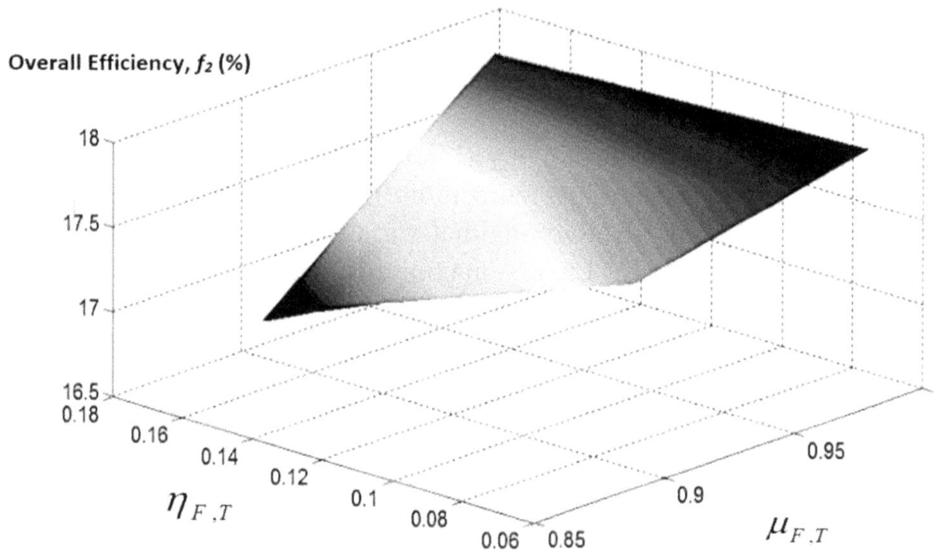

savings reaches the maximum of 144,113 USD (at $\mu_{F,T} = 0.9871$ and $\eta_{F,T} = 0.0665$) with the minimal of 141,434 USD ($\mu_{F,T} = 0.8967$ and $\eta_{F,T} = 0.1717$). The variation in fiscal savings with respect to the membership

Figure 15. Fiscal savings, f_3 (USD) relative too membership grades

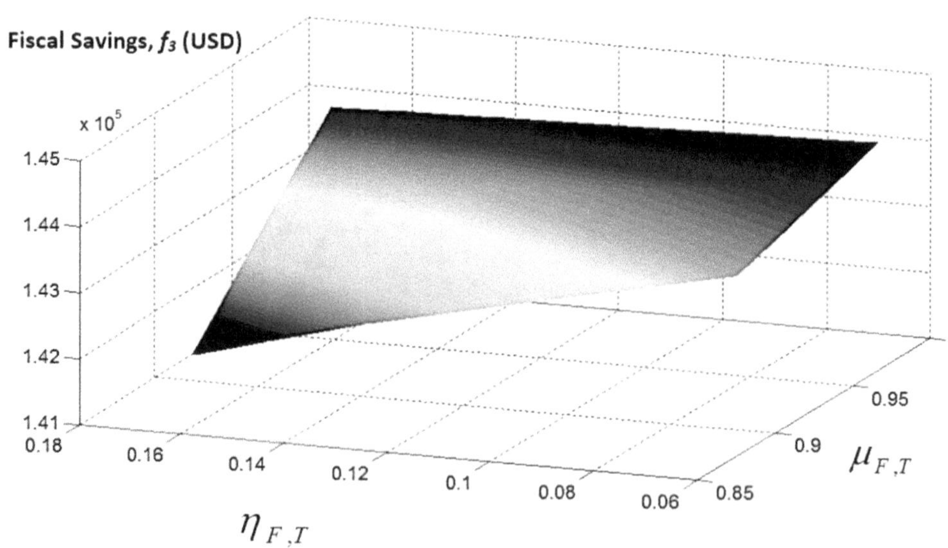

grades is 2679 USD. In Figures 13 – 15, it can be observed that except for the f_3 (fiscal savings), the variation in values of the objectives with respect to the membership grades are very small, (f_1, f_2, f_3) = (0.3671kW, 1.022%, 2679 USD). The design model presented in this chapter is for a single unit. In real-world applications, multiple units are conventionally utilized for stable power supply to large irrigation systems. Therefore, although the variations in the objectives are low, when projected to a larger scale such variations may compound producing a greater impact in terms of power output, system efficiency and fiscal savings. It should be noted, that even when considering a single-unit system, the variations in the membership grades (which spring from the noisy and uncertainty in insolation and ambient temperature) significantly affects the cost of the system (2679 USD).

For specific primary and secondary membership grades, three Pareto frontiers were selected based on the most optimal values of the objectives. The details on the membership grades are specified to construct these frontiers are given in Table 6.

Referring to Table 6, it is seen that the ambient temperature memberships are specified while the insolation memberships are left to vary in their ranges. This is done via the utilization of the symmetry properties in the membership grades. The individual solutions for various weights generated by the BFA were gauged and ranked based on the values of the aggregate objective function. Table 7 provides the ranked individual solutions for Frontier 1.

In Table 2.7, the best individual solution was obtained at the weights (0.1, 0.1, 0.8) while the worst solution was attained at (0.6, 0.3, 0.1). The weight assignment for the median solution was (0.3, 0.4, 0.3). The scattered solutions in the objective space approximating the Pareto Frontier 1 is shown in Figure 16:

The ranked individual solutions along with their respective aggregate objective functions and parameters for Frontier 2 is given in Table 8.

Table 6. Pareto frontiers and their membership grades

Description	$\mu_{F,T}$	$\eta_{F,T}$	$\mu_{F,S}$	$\eta_{F,S}$
Frontier 1	0.8967	0.17169	0.25264 - 0.39913	0.02907 - 0.92274
Frontier 2	0.9565	0.15725		
Frontier 3	0.9871	0.06648		

Table 7. Individual solution rankings for frontier 1

Description		Best	Median	Worst
Objective Function	f_1	20.6787	20.6666	20.6466
	f_2	17.1771	17.1173	17.0407
	f_3	148004	147917	147774
Decision Parameters	x_a	0.355846	0.354952	0.353833
	x_b	460.647	460.504	460.267
	x_c	677.284	674.784	671.6
	x_d	0.030827	0.030491	0.030037
Noise Factors	Z_a	251.838	251.837	251.835
	Z_b	328.974	328.972	328.97
Aggregate Objective	F	118407	44388.1	14794.9

The best, worst and median individual solutions presented in Table 2.8 are associated with the weights ((0.1, 0.1, 0.8), (0.1, 0.8, 0.1) and (0.1, 9.6, 0.3) respectively. The graphical representation of the Pareto frontier 2 is depicted in Figure 17.

Table 9 gives the ranked individual solutions for frontier 3 along with its respective parameters, noise factors and objective values.

Figure 16. Approximation of frontier 1

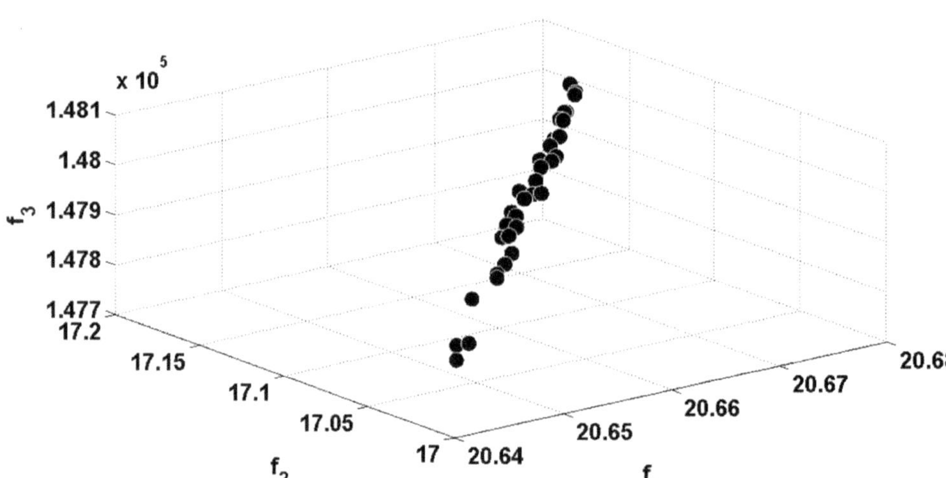

Table 8. Individual solution rankings for frontier 2

Description		Best	Median	Worst
Objective Function	f_1	20.9103	20.9103	20.8872
	f_2	17.7455	17.7506	17.6432
	f_3	149696	149696	149531
Decision Parameters	x_a	0.354758	0.354881	0.353615
	x_b	460.473	460.473	460.202
	x_c	674.513	674.732	670.244
	x_d	0.030421	0.030458	0.029886
Noise Factors	Z_a	277.854	277.854	277.852
	Z_b	328.972	328.972	328.969
Aggregate Objective	F	119761	44921.7	14969.3

The best and worst individual solutions in Table 9 are associated with the weights (0.1, 0.1, 0.8) and (0.5, 0.4, 0.1) respectively. The median solution has the weights (0.2, 0.5, 0.3). The Pareto Frontier 3 is shown in Figure 18:

The three frontiers generated by the BFA technique was obtained at various weights for three different membership grades (refer to Table 6). In this chapter, the overall dominance of the Pareto frontier is measured by taking the average value of the aggregate objective function, F across the entire frontier. Frontier 3 outranks Frontier 2 followed by Frontier 1. The average

Figure 17. Approximation of frontier 2

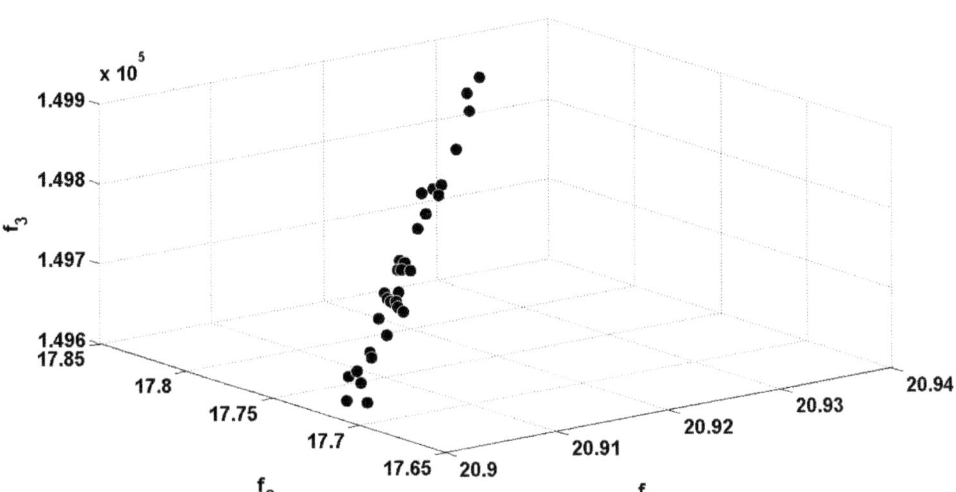

Table 9. Individual solution rankings for frontier 3

Description		Best	Median	Worst
Objective Function	f_1	21.034	21.0524	21.0335
	f_2	18.0657	18.1411	18.0577
	f_3	150600	150731	150596
Decision Parameters	x_a	0.354785	0.355775	0.354627
	x_b	460.435	460.65	460.428
	x_c	674.234	677.372	673.892
	x_d	0.030391	0.030815	0.030349
Noise Factors	Z_a	291.26	291.262	291.26
	Z_b	328.972	328.974	328.971
Aggregate Function	F	120484	45232.6	15077.3

value of the objective function across the entire frontier for Frontier 1, 2 and 3 are: 50,305.86, 50,911.64 and 51,237.71 respectively. Similarly in Tables 7 - 9, it can be observed that the value of the aggregate objective function for the best individual solution follows a similar trend as compared to the entire Pareto frontiers in terms of ranking. The Pareto frontier were also gauged using the SDM as presented in the previous section. The diversity levels obtained for each of the frontiers are presented in Figure 19:

In MO optimization, the capability of the solution technique to continuously produce solutions with high diversity is crucial. This is so that the solution

Figure 18. Approximation of frontier 3

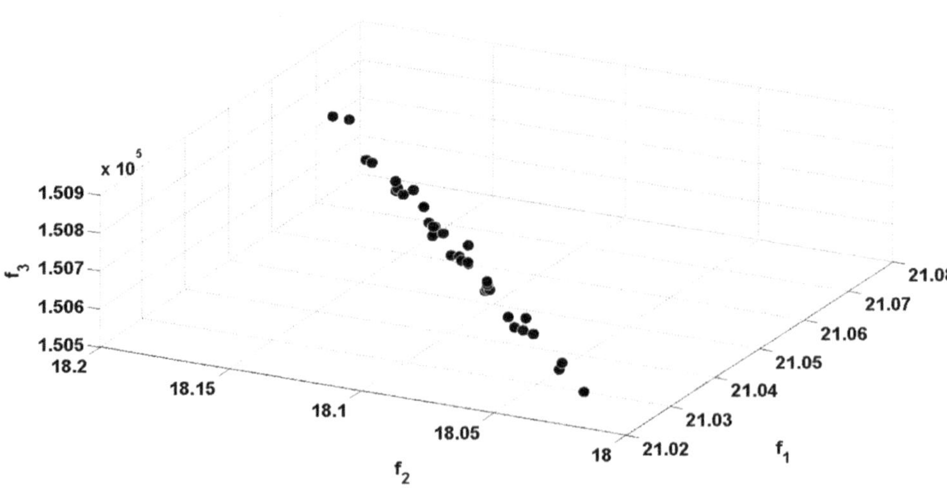

Figure 19. Diversity levels for the pareto frontiers

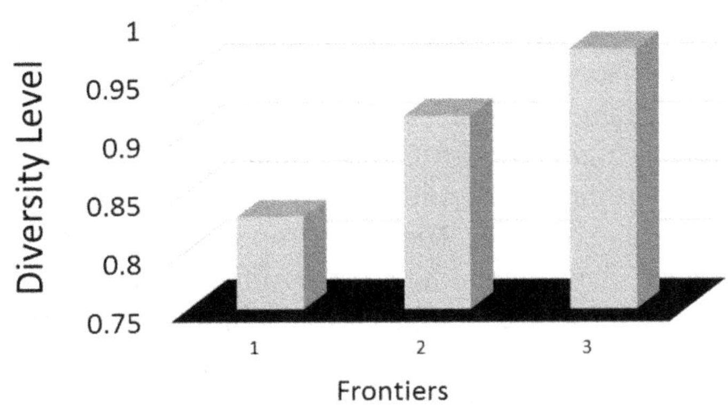

technique (algorithm) does not stagnate at certain locations in objective space or the local optima. Such stagnations may inhibit the algorithm from exploring other regions in the objective space which may contain other local optima which are much closer to the global optima. Therefore, measuring the diversity of the solution spread across the Pareto frontier provides the user/decision maker with information about the performance of the algorithm during its search operations. The BFA employed in this work seem to generate a highly diverse solution spread which constructs the Pareto Frontier 3 (Figure 18). Frontier 2 has lower diversity that Frontier 3 followed by Frontier 1 which has the lowest diversity as compared to all the frontiers generated using the BFA technique.

Since diverse solution generation provides the algorithm with the ability to overcome stagnation, the degree of frontier dominance closely relates to the diversity of the solution spread across the concerned frontier. In this work, it is seen that the diversity levels follow the ranking of the dominance measured by using the average aggregate objective function value. Therefore, the BFA approach generated diverse solutions during its execution which resulted in the construction of highly dominant Pareto frontiers.

One of the most vital issues in green energy engineering is the design of low-cost and efficient systems which could match (or surpass) its fossil-fuel counterparts. However, these design efforts often face challenges due to the system's interactions with noisy and uncertain environments. Identification and classification of the type of uncertainty is essential for the conception of suitable design frameworks. Using such frameworks, uncertainty-tolerant systems could be optimally designed and manufactured. In view of this idea, the data consisting of environmental factors presented in Table 5 could be seen to have two levels of division: monthly average data and annual average. Thus, modeling this data using a fuzzy type-1 system is clearly inadequate and may lead to inaccurate approximations to the solar-powered irrigation system's design characteristics. It could be observed that the reformulation of the problem in a fuzzy type-2 MO programming setting offers various new information such as the system tolerance limits and its behavior when under the influence of uncertainties. The three frontiers in Figures 16 - 18 were extracted because they contain the most significant variations in the target objectives when exposed to the uncertainties in the environmental factors. The maximal variation in the power output when faced with uncertainties (represented by the membership grades of the frontiers) is 0.3548 kW. The highest variation of the overall efficiency and the fiscal savings considering the mentioned uncertainties are 0.8806% and 2592 USD respectively. Therefore, in a large-scale solar irrigation system, uncertainties in the environmental factor (specifically: insolation levels and ambient temperature) does significantly affect the design properties of these systems.

The BFA technique like most metaheuristic approaches is a stochastic optimization technique. Therefore due to its random nature, multiple runs are required during execution for result consistency. In this work, each individual solution is obtained after 5 program executions which adds up to 175 runs per frontier. The parameter setting of the BFA were heuristically determined and were not varied in this work. The parameter setting for the BFA is as in Table 10:

Referring to Figure 10, it can be observed that the BFA has one overall loop plus four primary loops: chemotaxis, swimming, reproduction and elimination/dispersal. Due to its rigor in solution discrimination; portrayed by the cascaded loops, the BFA has high algorithmic complexity. The BFA code employed in this work contains multiple subroutines to account for the BFAs complexity. This significantly influences the computational time of the BFA during execution. The BFA program employed in this work takes an average of 4.217 seconds to compute each individual solution per run with an average of 1500 iterations. Therefore to construct the entire frontier for this application, the BFA takes approximately 12.3 minutes. Besides the algorithmic complexity of the BFA, the objective space of the solar-powered irrigation problem also contributes to the computational effort and time. This is because, the objective space in this problem contains multiple local optima that misleads the algorithm into stagnation. To overcome these

Table 10. Initial parameters for the BFA

Parameters	Values
Total Iteration, N_T	200
Population Size, P	25
Swimming Loop limit, N_S	5
Repellent Signal Width, W_{rep}	10
Attractant Signal Width, W_{att}	0.2
Repellent Signal Height, H_{rep}	0.1
Attractant Signal Height, H_{att}	0.1
Reproduction limit, N_r	5
Elimination limit, N_e	5

local optima and proceed with the computation, the BFA takes additional computational time.

Another important feature of a MO optimization approach is the fitness assignment. The fitness assignment in the computational approach affects the way the algorithm is driven during the search process. In this work, all three target objectives are maximized and no objectives are minimized. Therefore, the aggregate objective function is used as a fitness criterion for the BFA algorithm during the search. The fitness of swarm is considered improved if the aggregate objective function is maximized during the current iteration as compared to the previous iteration. It is important to note that although the diversity metric is employed to measure the diversity of the solutions in the frontier, the diversity metric was not assigned as a fitness criterion in the algorithm. The diversity metric was employed in an offline manner from time to time to ensure that the algorithm is producing a diverse population during execution.

Diversity preservation is crucial when it comes to dealing with MO programming algorithms. In this work, two diversity preservation mechanisms were utilized. The first one as mentioned previously is the offline implementation of the diversity metric. The second approach is the in-built diversification mechanism of the BFA. The in-built mechanism of the BFA consists of the two subcomponents which are the initial randomization and the random dispersal components (see Figure 10). The initial randomization component randomly locates each of the bacterium at various locations in the objective space prior to the search. The random dispersal component, spreads the bacteria which survived elimination randomly throughout the objective space. This random spreading diversifies the bacteria population preparing it for the next program iteration. These two subcomponents randomizes the BFA during execution such that the bacterium population is sufficiently diverse during the search.

In the BFA, solution elitism could be readily ensured by the process of elimination (refer to Figure 10). In this process, bacteria which does not perform improvements based on the fitness assignment (which in this case is the maximization of the aggregate objective function) is eliminated from the population pool. This elitism mechanism ensures that only the elite bacteria is allowed to swarm during the consequent program cycles. Throughout the execution, the BFA algorithm performed in a stable manner and managed to successfully converge to a solution during each run. Besides, none of the fuzzy constraints were violated by solutions generated by the BFA. Therefore, the BFA reliably produces feasible solutions for all the associated weights. This in effect enables the BFA to capture solutions in the objective space which are close to the global optima.

FINAL REMARKS

In Part 1, a new local maximum and an efficient construction of the Pareto frontier were achieved using the Hyp-AP approach. The HVI was used to gauge the individual solutions produced by the w-AP, Hyp-AP and the DSIDES (Chen *et al.*, 1995) algorithms. On the other hand in Part 2, the generalized MO design of the solar-powered irrigation system was completed using type-2 fuzzy modeling to represent uncertain environmental factors. Using the BFA in tandem with the weighted-sum approach, dominant Pareto frontiers were constructed using individual solutions (with various weights). The dominance of the three generated Pareto frontiers were evaluated and ranked. The design problem was successfully reformulated using type-2 fuzzy logic to depict uncertainty in the environmental data (as shown in Table 5). As seen in Tables 7 - 9, new optimal individual solutions were obtained. In addition, information regarding the effects of uncertainties on the variations in the target objectives were ascertained and analyzed. This knowledge would prove to be very useful for designers dealing with solar-powered irrigation systems which operate in uncertain (or noisy) environments. In addition, using the diversity metric and the aggregate objective function, the quality of the solutions as well as the behavior of the algorithm was analyzed in detail. The operations and the mechanisms of the BFA related to the solar-powered irrigation problem was presented. Characteristics such as the algorithmic complexity and execution time was also presented and discussed.

It can be concluded that although the BFA consumes high computational resources, it compensates in terms rigor in solution scrutiny. This in turn provides the engineer/decision maker with high quality solutions. Based on the results obtained in this work, it can be seen that if a multi-unit (large-scale) solar-powered irrigation system were to designed for operation in an uncertain environment, its key properties such as efficiency, cost and power output would be greatly affected. The following are some ideas to improve the efficiency of the techniques presented in this chapter:

- One method to address the issue of algorithmic uncertainty is to vary the initial parameter setting of the BFA. This way, the optimal parameters could be identified for solving a particular design problem with uncertain environmental factors.
- Other conventional fuzzy-based approaches (Vasant *et al.*, 2010; Ganesan *et al.*, 2014) could be upgraded to a type-2 fuzzy framework and implemented to this problem.
- Alternative forms of metaheuristic approaches such using evolutionary strategies could be implemented to this design problem (Ganesan *et al.*, 2015).
- Different measurement metrics could also be employed for gauging the degree of dominance of the generated Pareto frontiers (Zitzler and Thiele, 1998).
- A study to rigorously determine the fuzzy parameters (equation (2.9)) could help obtain better modeling accuracy - this could potentially be would be very useful for future researchers attempting to optimize the design of systems plagued with high levels of uncertainties as encountered in this chapter .

REFERENCES

Al-Ali, A. R., Rehman, S., Al-Agili, S., Al-Omari, M. H., & Al-Fayezi, M. (2001). Usage of photovoltaic in an automated irrigation system. *Renewable Energy*, *23*(1), 17–26. doi:10.1016/S0960-1481(00)00110-5

Al-Hadi, I. A. A., & Hashim, S. I. M. (2011). Bacterial Foraging Optimization Algorithm For Neural Network Learning Enhancement. *International Journal of Innovative Computing*, *1*(1), 8–14.

Allawi, Z. T. (2014). A PSO-optimized type-2 fuzzy logic controller for navigation of multiple mobile robots. *19th International Conference on Methods and Models in Automation and Robotics (MMAR)*, 33 - 39. 10.1109/MMAR.2014.6957321

Bahraminejad, B., Iranpour, M. R., & Esfandiari, E. (2014). Pitch Control of Wind Turbines Using IT2FL Controller versus T1FL Controller. *International Journal Of Renewable Energy Research*, *4*(4), 1065–1077.

Beume, N., Naujoks, B., & Emmerich, M. (2007). SMS-EMOA: Multiobjective selection based on dominated hypervolume. *European Journal of Operational Research*, *181*(3), 1653–1669. doi:10.1016/j.ejor.2006.08.008

Castillo, O., & Melin, P. (2012). A review on the design and optimization of interval type-2 fuzzy controllers. *Applied Soft Computing*, *12*(4), 1267–1278. doi:10.1016/j.asoc.2011.12.010

Chen, W., Kwok-Leung, T., Allen, J. K., & Mistree, F. (1995). *Integration of the Response Surface Methodology with the compromise decision support problem in developing a general robust design procedure. In Advances in Design Automation* (Vol. 82-2). New York: ASME.

Das, I., & Dennis, J. E. (1998). Normal-boundary intersection: A new method for generating the Pareto surface in nonlinear multicriteria optimization problems. *SIAM Journal on Optimization*, *8*(3), 631–657. doi:10.1137/S1052623496307510

Dereli, T., Baykasoglu, A., Altun, K., Durmusoglu, A., & Türksen, I. B. (2011). Industrial applications of type-2 fuzzy sets and systems: A concise review. *Computers in Industry*, *62*(2), 125–137. doi:10.1016/j.compind.2010.10.006

Dhimmar, V., Prajapti, J., Patel, M., Patel, D., Mistry, B., & Parmar, J. (2014). Design of Solar Steam Irrigation Pump. *International Journal of Engine Research*, *3*(5), 315–317. doi:10.17950/ijer/v3s5/504

El-Wahed, W. F. A., Mousa, A. A., & Elsisy, M. A. (2008). Solving Economic Emissions Load Dispatch problem by using Hybrid ACO-MSM approach. *The Online Journal on Power and Energy Engineering*, *1*, 31–35.

Elamvazuthi, Ganesan, & Vasant. (2011). A Comparative Study of HNN and Hybrid HNN-PSO Techniques in the Optimization of Distributed Generation (DG) Power Systems. *International Conference on Advance Computer Science and Information System*, 195-199.

Fayek, H. M., Elamvazuthi, I., Perumal, N., & Venkatesh, B. (2014). A controller based on Optimal Type-2 Fuzzy Logic: Systematic design, optimization and real-time implementation. *ISA Transactions*, *53*(5), 1583–1591. doi:10.1016/j.isatra.2014.06.001 PMID:24962934

Ganesan, T., Elamvazuthi, I., Shaari, K., & Vasant, P. (2012). Swarm intelligence and gravitational search algorithm for multi-objective optimization of synthesis gas production. *Journal of Applied Energy*. . doi:10.1016/j.apenergy.2012.09.059

Ganesan, T., Elamvazuthi, I., Shaari, K. Z. K., & Vasant, P. (2013). Hypervolume-driven analytical programming for solar-powered irrigation system optimization. In *Nostradamus 2013: Prediction, Modeling and Analysis of Complex Systems* (pp. 147–154). Heidelberg, Germany: Springer. doi:10.1007/978-3-319-00542-3_15

Ganesan, T., Elamvazuthi, I., & Vasant, P. (2015). Multiobjective Design Optimization of a Nano-CMOS Voltage-Controlled Oscillator Using Game Theoretic-Differential Evolution. *Applied Soft Computing*, *32*, 293–299. doi:10.1016/j.asoc.2015.03.016

Ganesan, T., Vasant, P., & Elamvazuthi, I. (2014). Hopfield neural networks approach for design optimization of hybrid power systems with multiple renewable energy sources in a fuzzy environment. *Journal of Intelligent & Fuzzy Systems*, *26*(5), 2143–2154.

Ganesan, T., Vasant, P., & Elamvazuthi, I. (2017). Multiobjective Optimization of Solar-Powered Irrigation System with Fuzzy Type-2 Noise Modelling. In Emerging Research on Applied Fuzzy Sets and Intuitionistic Fuzzy Matrices (pp. 189-214). IGI Global. doi:10.4018/978-1-5225-0914-1.ch008

Gouws, R., & Lukhwareni, T. (2012). Factors influencing the performance and efficiency of solar water pumping systems: A Review. *International Journal of Physical Sciences*, *7*(48), 6169–6180.

Greenfield, S., John, R., & Coupland, S. (2005). A novel sampling method for type-2 defuzzification. *Proc. UKCI 2005*, 120–127.

Hamrawi, H., & Coupland, S. (2009). Type-2 fuzzy arithmetic using Alpha-planes. *Proceedings of the Joint 2009 International Fuzzy Systems Association World Congress and 2009 European Society of Fuzzy Logic and Technology conference*, 606-611.

Helikson, H. J., Haman, D. Z., & Baird, C. D. (1991). *Pumping water for irrigation using solar energy, Fact Sheet (EES-63)*. University of Florida.

Holland, J. H. (1992). *Adaptation in Natural and Artificial Systems: An Introductory Analysis with Applications to Biology, Control and Artificial Intelligence*. MIT Press. doi:10.7551/mitpress/1090.001.0001

Igel, C., Hansen, N., & Roth, S. (2007). Covariance matrix adaptation for multi-objective optimization. *Evolutionary Computation*, *15*(1), 1–28. doi:10.1162/evco.2007.15.1.1 PMID:17388777

Jasim, K. K., Kadhum, J. A., & Ali, M. (2014). Design and Construction of Hybrid Solar-Wind System used for Irrigation Projects. *Indian Journal of Applied Research*, *4*(7), 518–522. doi:10.15373/2249555X/July2014/163

Kennedy, J., & Eberhart, R. (1995). Particle Swarm Optimization. *IEEE Proceedings of the International Conference on Neural Networks,* 1942-1948.

Klir, G. J., & Yuan, B. (1995). *Fuzzy Sets and Fuzzy Logic: Theory and Applications*. Upper Saddle River, NJ: Prentice Hall.

Knowles, J., & Corne, D. (2003). Properties of an adaptive archiving algorithm for storing nondominated vectors. *IEEE Transactions on Evolutionary Computation*, *7*(2), 100–116. doi:10.1109/TEVC.2003.810755

Koza, J. R. (1992). *Genetic Programming: On the Programming of Computers by means of Natural Selection*. MIT Press.

Linda, O., & Manic, M. (2011). Evaluating Uncertainty Resiliency of Type-2 Fuzzy Logic Controllers for Parallel Delta Robot. *4th International Conference on Human System Interactions (HSI)*, 91 - 97. 10.1109/HSI.2011.5937349

Liu, F. (2006). *An efficient centroid type reduction strategy for general type-2 fuzzy logic systems. Walter J. Karplus Summer Research Grant Report*. IEEE Computational Intelligence Society.

Liu, Y., Passino, K. M., & Simaan, M. A. (2002). Biomimicry of social foraging bacteria for distributed optimization: Models, principles, and emergent behaviors. *Journal of Optimization Theory and Applications*, *115*(3), 603–628. doi:10.1023/A:1021207331209

Lucas, L., Centeno, T., & Delgado, M. (2007). General type-2 fuzzy inference systems: Analysis, design and computational aspects. *Proceedings of IEEE International Conference of Fuzzy Systems*, 1107–1112. 10.1109/FUZZY.2007.4295522

Martinez, R., Castillo, O., Aguilar, L. T., & Rodriguez, A. (2011). Optimization of Type-2 Fuzzy Logic Controllers Using PSO Applied to Linear Plants, Soft Computing for Intelligent Control and Mobile Robotics. *Studies in Computational Intelligence*, *318*, 181–193.

Masand, B. (1994). *Optimizing confidence of text classification by evolution of symbolic expressions. In Advances in Genetic Programming*. MIT Press.

Mendel, J., & John, R. (2002). Type-2 fuzzy sets made simple. *IEEE Transactions on Fuzzy Systems*, *10*(2), 117–127. doi:10.1109/91.995115

Mezura-Montes, E., & Hern'andez-Ocana, B. (2009). Modified bacterial foraging optimization for engineering design. *Proceedings of the Artificial Neural Networks in Engineering Conference (ANNIE'2009)*, 357–364.

Mezura-Montes, E., Portilla-Flores, E. A., & Hern'andez-Oca˜na, B. (2014). Optimum synthesis of a four-bar mechanism using the modified bacterial foraging algorithm. *International Journal of Systems Science*, *45*(5), 1080–1100. doi:10.1080/00207721.2012.745023

Mistree, F., Hughes, O. F., & Bras, B. A. (1993). *The Compromise Decision Support Problem and the Adaptive Linear Programming Algorithm. In Structural Optimization: Status and Promise* (pp. 247–286). Washington, DC: AIAA.

Mo, H., Wang, F. Y., Zhou, M., Li, R., & Xiao, Z. (2014). Footprint of uncertainty for type-2 fuzzy sets. *Information Sciences*, *272*, 96–110. doi:10.1016/j.ins.2014.02.092

Mostaghim, S., & Teich, J. (2003). Strategies for finding good local guides in multiobjective particle swarm optimization. *IEEE Swarm Intelligence Symposium*, 26-33.

Mostaghim, S., & Teich, J. (2005). A New Approach on Many Objective Diversity Measurement. *Dagstuhl Seminar Proceedings 04461, Practical Approaches to Multi-Objective Optimization*, 1-15.

Nouri, H., & Hong, T. S. (2013). Development of bacteria foraging optimization algorithm for cell formation in cellular manufacturing system considering cell load variations. *Journal of Manufacturing Systems*, *32*(1), 20–31. doi:10.1016/j.jmsy.2012.07.014

Nurcahyo, G. W. (2003). Selection of Defuzzification Method to Obtain Crisp Value for Representing Uncertain Data in a Modified Sweep Algorithm. JCS&T, 3(2).

O'Brien, W., Browman, H., & Evans, B. (1990). Search strategies of foraging animals. *American Scientist*, *78*, 152–160.

Oakley, H. (1994). Two scientific applications of genetic programming: Stack filters and non-linear equation fitting to chaotic data. In *Advances in Genetic Programming*. MIT Press.

Oh, S. K., Jang, H. J., & Pedrycz, W. (2011). A comparative experimental study of type-1/type-2 fuzzy cascade controller based on genetic algorithms and particle swarm optimization. *Expert Systems with Applications*, *38*(9), 11217–11229. doi:10.1016/j.eswa.2011.02.169

Ontoseno, P., Abdillah, M., Wibowo, R., & Soeprijanto, A. (2013). Optimal Design of Photovoltaic–Battery Systems Using Interval Type-2 Fuzzy Adaptive Genetic Algorithm. *Engineering*, *5*(1), 50–55. doi:10.4236/eng.2013.51B009

Oplatkova, Z., & Zelinka, I. (2006). Investigations using Artificial Ants using Analytical Programming. *Second International Conference on Computational Intelligence, Robotics, and Autonomous Systems.*

Oplatkova, Z., & Zelinka, I. (2007). Creating evolutionary algorithms by means of analytic programming - design of new cost function. In *ECMS 2007* (pp. 271–276). European Council for Modelling and Simulation. doi:10.7148/2007-0271

Panda, S. K., Padhee, S., Kumar Sood, A., & Mahapatra, S. S. (2009). Optimization of Fused Deposition Modelling (FDM) Process Parameters Using Bacterial Foraging Technique, *Intelligent. Information & Management*, *1*, 89–97.

Paricheh, M., & Zare, A. (2013). Traffic Flow Prediction Based on Optimized Type-2 Neuro-Fuzzy Systems. *International Journal of Engineering and Computer Science*, *2*(8), 2434–2439.

Passino, K. (2002). Biomimicry of bacterial foraging for distributed optimization and control. *IEEE Control Systems Magazine*, *22*(3), 52–67. doi:10.1109/MCS.2002.1004010

Pytel, K. (2012). The fuzzy Genetic System for Multiobjective Optimization. *Proceedings of the Federated Conference on Computer Science and Information Systems*, 137 - 140. 10.1007/978-3-642-29353-5_38

Qiu, C., Xiao, J., Yu, L., Han, L., & Iqbal, M. N. (2013). A modified interval type-2 fuzzy C-means algorithm with application in MR image segmentation. *Journal of Pattern Recognition Letters*, *34*(12), 1329–1338. doi:10.1016/j.patrec.2013.04.021

Rao, D. H., & Saraf, S. S. (1996). Study of defuzzification methods of fuzzy logic controller for speed control of a DC motor. *Proceedings of the 1996 International Conference on Power Electronics, Drives and Energy Systems for Industrial Growth*, 2, 782 - 787. 10.1109/PEDES.1996.535878

Ravi, K., Shilaja, C., Chitti Babu, B., & Kothari, D. P. (2014). Solving Optimal Power Flow Using Modified Bacterial Foraging Algorithm Considering FACTS Devices. *Journal of Power and Energy Engineering*, *2*(04), 639–646. doi:10.4236/jpee.2014.24086

Rommelfanger, H. (1989). Interactive decision making in fuzzy linear optimization problems. *European Journal of Operational Research, 41*(2), 210–217.

Ryan, C., Collins, J., & O'Neill, M. (1998). Grammatical evolution: Evolving programs for an arbitrary language. Lecture Notes in Computer Science, *First European Workshop on Genetic Programming*. 10.1007/BFb0055930

Sánchez, D., Melin, P., & Castillo, O. (2015). Fuzzy System Optimization Using a Hierarchical Genetic Algorithm Applied to Pattern Recognition. *Advances in Intelligent Systems and Computing*, *323*, 713–720. doi:10.1007/978-3-319-11310-4_62

Schneider, S. S., & McNally, L. C. (1992). Colony Defense in the African Honey Bee In Africa. *Environmental Entomology*, *21*(6), 1362–1370. doi:10.1093/ee/21.6.1362

Turner, J. S. (2011). Termites as models of swarm cognition. *Swarm Intelligence*, *5*(1), 19–43. doi:10.100711721-010-0049-1

Varacha, P. (2011). *Neural Network Synthesis via Asynchronous Analytical Programming*. Recent Researches in Neural Networks, Fuzzy Systems, Evolutionary Computing and Automation.

Vasant, P., Elamvazuthi, I., Ganesan, T., & Webb, J. F. (2010). Iterative fuzzy optimization approach for crude oil refinery industry. *Scientific Annals of Computer Science*, *8*(2), 261–280.

Vijay, R. (2012). Intelligent Bacterial Foraging Optimization Technique to Economic Load Dispatch Problem. *International Journal of Soft Computing and Engineering*, *1*(2), 2231–2307.

Wong, Y. W., & Sumathy, K. (2001). Thermodynamic analysis and optimization of a solar thermal water pump. *Applied Thermal Engineering*, *21*(5), 613–627. doi:10.1016/S1359-4311(00)00065-X

Wu, D., & Tan, W. W. (2004). A type-2 fuzzy logic controller for the liquid-level process. *Proceedings of 2004 IEEE International Conference on Fuzzy Systems*, 2, 953 - 958. 10.1109/FUZZY.2004.1375536

Yildiz, A. R. (2013). Cuckoo search algorithm for the selection of optimal machining parameters in milling operations. *International Journal of Advanced Manufacturing Technology*, *64*(1-4), 55–61. doi:10.100700170-012-4013-7

Zadeh, L. (1975). The concept of a linguistic variable and its application to approximate reasoning—I. *Information Sciences*, *8*(3), 199–249. doi:10.1016/0020-0255(75)90036-5

Zelinka, I. (2002). Analytic programming by Means of Soma Algorithm. *Proceeding of the 8th International Conference on Soft Computing Mendel'02*, 93-101.

Zelinka, I., & Oplatkova, Z. (2003). Analytic programming – Comparative Study. CIRAS'03. *Second International Conference on Computational Intelligence, Robotics, and Autonomous Systems*.

Zitzler, E., & Thiele, L. (1998). Multiobjective Optimization Using Evolutionary Algorithms - A Comparative Case Study. *Conference on Parallel Problem Solving from Nature (PPSN V)*, 292–301. 10.1007/BFb0056872

Zitzler, E., & Thiele, L. (1999). Multiobjective evolutionary algorithms: A comparative case study and the strength Pareto approach. *IEEE Transactions on Evolutionary Computation*, *3*(4), 257–271. doi:10.1109/4235.797969

Chapter 3
Multiobjective Programming for Waste Heat Recovery of an Industrial Gas Turbine

ABSTRACT

In gas power plants, the overall efficiency of the generation system plays a key role in ensuring stable and efficient power supply. Terms and conditions of power supply are usually detailed in power purchase agreements (PPA). Current requirements set by PPAs limit the net power produced by the supplier. This creates opportunities for plant optimization efforts to focus on system efficiency—aiming to increase system lifetime with lower operational costs. In this chapter, a gas turbine (GT) system is considered to demonstrate certain features of power plant optimization. Waste heat from the GT exhaust stack is fed into an absorption chiller (AC). The AC cools the air intake at the GT compressor. This cooling reduces the heat rate and increases the GT efficiency. This combined GT-AC system was optimized in a multi-objective (MO) setting while considering power limitations (imposed by the PPA).

DOI: 10.4018/978-1-7998-1710-9.ch003

PART 1: OPTIMIZATION OF THE THREE-OBJECTIVE GT-AC SYSTEM WITHOUT UNCERTAINTY IN WEATHER CONDITIONS

Overview

In most power production industries, the overall efficiency of the generation system plays a key role. Optimizing the system increases its efficiency resulting in higher power production. Most power producers are bound contractually by PPAs which details the amount of power they are allowed to produce (Wu and Babich, 2012; Cory *et al.*, 2009). The strategies involved in drawing up the contract vary from country to country and represent an alignment towards local policies and scenarios. Some of these contracts require that the amount of power supplied remain fixed throughout the period of purchase. Nevertheless, there may be requirements specified in the contracts to combine fixed generation with additional available power. This is done to address concerns over unplanned outages. For contracts (PPA) involving fixed power supply: although engineers and plant personnel manage to optimize the plant efficiency significantly, the surplus power produced could not be sold. The motivation to pursue power generation efficiency in this sense can also relate to savings in fuel consumption - without the need to increase the power supplied. This work aims to optimize the design and operation parameters for the combined Gas Turbine (GT) and Absorption Chiller (AC) system without maximizing the total power supplied to the consumer. The optimization model for the combined system would hence contain multiple objectives (which represent key indicators of efficiency) while constraining the GT power output such that it adheres to PPA requirements.

By optimizing the design and operations of the combined GT-AC system, the cost efficiency of the plant would significantly increase. Additionally, this optimization would also enable the GT to operate at maximum efficiency as per design. This would contribute to the increase of its active lifetime. Recovering the waste heat works in favor of the environment since the plant does not dispose large amounts of heat to its surrounding.

Real-world engineering applications often present scenarios with multiple target objectives. Such classes of problems are often difficult to solve. Even if solved, these solutions are difficult to analyze due to their multidimensional and complex nature. However, as industrial systems become more complex, engineers and decision makers often find themselves in situations involving multiple objectives (Ganesan *et al.*, 2014; Ganesan *et al.*, 2015). The idea of Pareto-optimality is prevalent when it comes to tracing-out the non-dominated solution options at the Pareto curve (Deb *et al.*, 2002). An alternative to non-dominated solution tracing is the function aggregation method – where multiple objective functions are merged into a single master function effectively transforming the problem into a weighted single-objective problem (Marler and Arora, 2010; Brito *et al.*, 2014; Naidu *et al.*, 2014). Detail examples and analyses on MO techniques for problems in engineering optimization are presented in Oliveira and Saramago, (2010) and Rao and Rao, (2009).

MO scenarios have been encountered in power generation and aerospace - especially when dealing with gas turbines (GT). For instance, in García-Revillo *et al.*, (2014), a Multiobjective Genetic Algorithm (MOGA) was employed for optimizing the geometry of aeronautical GT discs. In that work the authors considered fatigue life prediction and total geometrical mass as objective functions. In Yazdi, *et al.*, (2015), the modeling and optimization of a micro turbine cycle was done – where the design parameters; compression ratio, compressor isentropic efficiency, combustion chamber inlet temperature and turbine inlet temperature was considered. The authors considered power exergy efficiency, total cost and carbon dioxide emission of the plant as the three objectives to be optimized. Khorasani Nejad *et al.*, (2013) successfully optimized the total cost rate and power cycle efficiency of a GT power plant coupled with an AC system (which cools the compressor inlet air). The authors employed a Multiobjective Genetic Algorithm as the optimization technique. For accurately describing a combined system encompassing a GT, heat recovery steam generator and a LiBr-based AC, a multiobjective evaluation index (MEI) model was developed by Sun *et al.*, (2014). The proposed model was then utilized to optimize the combined system. A similar MO combined system was modeled in Ahmadi *et al.*, (2013). Their system consisted of a micro gas turbine, boiler, AC, ejector refrigeration cycle, domestic water heater and a proton-exchange membrane electrolyzer. The system in Ahmadi *et al.*, (2013) was aimed to produce power, heating, cooling, hot water and hydrogen. Using a non-dominated sorting genetic algorithm (NSGA-II), the authors optimized the combined system focusing on two objectives; total cost rate and system's exergy efficiency.

Evolutionary algorithms have proven to be highly successful when used for solving problems in engineering optimization. Among the most successful evolutionary techniques are differential evolution (DE) (Price *et al.*,2006), genetic algorithms (Grefenstette, 2013), evolutionary strategy (Beyer, 2013) and genetic programming (Shao *et al.*, 2014). DE is a population-based evolutionary algorithm that has been derived from genetic algorithms. Developed in the nineties, DE has been successfully applied to engineering problems which are non-differentiable, non-continuous, non-linear, noisy, and multidimensional. These problems often contain many local minima, constraints and have a high degree of stochasticity. Lately, DE has been applied to a variety of areas including optimization problems in energy systems (Rout, 2013; Mohanty *et al.*, 2014; Debbarma *et al.*, 2014). In this work the basic DE technique is employed in conjunction to the weighted sum approach (Liu *et al.*, 2011). To obtain a solution closer to the global optima, the random generator of the conventional DE technique is enhanced with a chaotic component. As previously mentioned solution evaluation for MO problems could prove difficult. Therefore in this work the Hypervolume Indicator (HVI) is employed for this purpose (Zitzler *et al.*,2007). The HVI is a set measure reflecting the volume enclosed by a Pareto front approximation and a reference set (Emmerich *et al.*, 2005; Jiang *et al.*, 2015).

This primary focus of this chapter is the novel modeling of a real-world combined GT and AC multicriteria optimization problem. Key concepts of the model as well as the solution method presented in this chapter can be found in the author's original past works (Ganesan *et al.*, 2016; Ganesan *et al.*, 2018). This model is optimized by maximizing the thermal efficiency of the GT system and the coefficient of performance (COP) of the AC. Optimizing the fuel consumption, the fuel cost is minimized while maintaining constant power supply to respect the PPA with the consumer. To obtain optimal parameters, two evolutionary-type algorithms are employed: DE and the improved CDDE. Optimization results are discussed and analyzed in detail in Part 1 of this chapter.

POWER PURCHASE AGREEMENTS

Most electrical power generation plants are driven by fossil fuel or renewable energy sources. To ensure realistic returns and financial security, power suppliers usually draw-up a contract with the client(s) - called Power Purchase Agreement (PPA). The PPA is a principal contract between two parties: the electricity provider (seller) and the purchaser (the buyer/client) (Javadi and Javadinasab, 2011). PPAs are usually long-term agreements lasting between 5 and 20 years. Independent power producers are the common power suppliers under PPAs. Today, many variants of PPAs exist differing according to the necessities of the buyer, supplier and their financing counterparts.

PPAs remain essential for power purchasers since it provides assurance in terms of security of power supply. Besides it also offers protection from competition with other producers - which offer cheaper power sales. Feasibility of power engineering projects (e.g. cost of building the power plant and transmission infrastructure) could be determined by considering the terms in PPAs. Commercial terms between the two parties such as commercial operation date, schedule for transfer of electricity, penalties for under-delivery, payment terms and also termination conditions are clearly defined in the PPA.

Scheduled maintenance outages, operations, maintenance, emergencies, accounting and record keeping practices are also incorporated in the PPA (Bierdel, 2013). PPAs are used by most countries to regulate the sales and purchase of power. For example, in Indonesia PPAs were introduced in 1991 when private participation in power generation was permitted (Negara, 2013). In early 2015, giant companies such as Google and Apple in the United States also participated in signing renewable energy PPAs (Labrador, 2015). In Malaysia, PPAs were introduced in 1993 when the government permitted the private sector to supply the required power generating capacity.

PPAs impose an upper limit on the power that the supplier can produce for the client according to specific electricity rates. If the seller produces a surplus of energy, there will be a negative effect on the sales electricity rates. Thus to optimize the power plant, the supplier needs to look at other engineering aspects without increasing the power output from the power plant. This way the PPAs would be respected and the power supply is optimized. Amongst the power generation plants bound by PPAs, the GT plant is the most versatile and flexible in its generation response. A GT plant can deliver its target output within minutes of a power request as compared to other thermal plants. One potentially effective optimization approach is to increase the thermal efficiency of the GT and reducing its heat rate while maintaining constant power supply. This way the overall fuel consumption of the GT could be minimized and power could be generated in a more efficient, cheaper and environmentally friendly manner. This increase in efficiency could contribute to the increase of GT engine lifetime since it operates at its optimal setting.

DYNAMIC MODEL

The model developed in this work is of a combined GT and AC system – where the AC is used to cool the compressor inlet air of the GT. The AC system uses LiBr as a refrigerant. This cooling is aimed to enhance the performance of the GT in terms of: overall thermal efficiency (η_{th}) as well as lowering the heat rate (HR). The GT simulation in this work was modeled based on a V94.2 Siemens GT with a rated speed of 3000 rpm and a rated capacity of 131.5 MW. The GT in the plant as well as in the simulation were assumed to be operated at a base load of 123 MW (note that although the case employed in this chapter was based on an industrial-scale plant, the data employed has been intentionally modified for confidentiality purposes). The seven day hourly-averaged ambient temperature profile for a day considered in this work is depicted in Figure 1:

The heat exchanger design, pumps and compressor formulations were neglected in this model since they have very minimal effect on the overall optimization. The formulations for the concentration of strong and weak LiBr solution (X_S and X_W) are as follows:

$$X_S = \frac{49.04 + 1.125t_a - t_e}{134.65 + 0.47t_a} \text{ where } X_S = X_1 = X_2 = X_3 \tag{3.1}$$

Figure 1. Daily - average ambient temperature

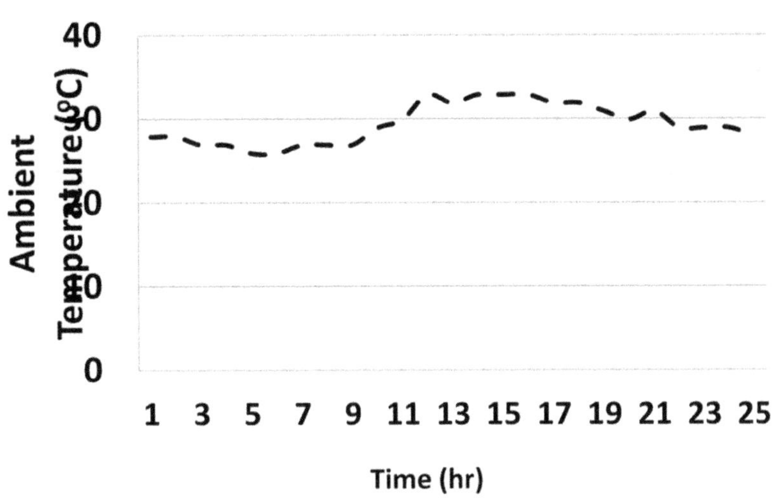

$$X_W = \frac{49.04 + 1.125t_g - t_c}{134.65 + 0.47t_g} \text{ where } X_W = X_4 = X_5 = X_6 \tag{3.2}$$

For pure water, the concentrations zero off as follows:

$$X_7 = X_8 = X_9 = X_{10} = 0 \tag{3.3}$$

The pressures for the evaporator and condenser components in the AC are as follows:

$$\log_{10} P_e = 7.8553 - \frac{1555}{t_e + 273.15} - \frac{11.2414 \times 10^4}{(t_e + 273.15)^2} \text{ where } P_e = P_1 = P_6 = P_9 = P_{10} \tag{3.4}$$

$$\log_{10} P_c = 7.8553 - \frac{1555}{t_c + 273.15} - \frac{11.2414 \times 10^4}{(t_c + 273.15)^2} \text{ where } P_c = P_2 = P_3 = P_4 = P_5 = P_7 = P_8 \tag{3.5}$$

Considering the throttling effect where $H_5 = H_6$ and $H_8 = H_9$, the enthalpies are as follows:

$$H_8 = H^l_{H_2O} = t_c - 25 \tag{3.6}$$

$$H_{10} = 572.8 + 0.417t_e \tag{3.7}$$

Transient refrigeration load and its corresponding mass balance formulations are:

$$\frac{dQ_E}{dt} = \dot{m}_R(H_{10} - H_9) \tag{3.8}$$

$$\dot{m}_W X_6 + \dot{m}_R X_{10} = \dot{m}_S X_1 = \left(\dot{m}_W + \dot{m}_R\right) X_1 \tag{3.9}$$

$$\therefore m_S = \dot{m}_W + \dot{m}_R \tag{3.10}$$

$$\dot{m}_W = \left(\frac{Q_E}{H_{10} - H_8}\right)\left(\frac{X_1}{X_4 - X_1}\right) \tag{3.11}$$

$$\dot{m}_S = \left(\frac{Q_E}{H_{10} - H_8}\right)\left(\frac{X_4}{X_4 - X_1}\right) \tag{3.12}$$

Since the generator and absorbers have sumps and tubes, some fluid retention occurs in these components. The resulting time delay is modeled according to the formulations built in Kohlenbach and Ziegler, (2008). The tube mass flowrate at the absorber and generator are as follows:

$$\dot{m}_{a,t}(i) = \dot{m}_S + \left[\frac{M_a(i) - M_a(i-1)}{\Delta t}\right] \tag{3.13}$$

$$\dot{m}_{g,t}(i) = \dot{m}_W + \left[\frac{M_g(i) - M_g(i-1)}{\Delta t}\right] \tag{3.14}$$

The specific heats and inlet temperatures at the solution heat exchangers are:

$$C_{X_1} = 1.01 - 1.23X_1 + 0.48X_1^2 \tag{3.15}$$

$$C_{X_4} = 1.01 - 1.23X_4 + 0.48X_4^2 \tag{3.16}$$

$$t_5 = t_g - E_L(t_g - t_a) \tag{3.17}$$

$$t_3 = t_a - E_L \left(\frac{X_1}{X_4}\right)\left(\frac{C_{X_4}}{C_{X_1}}\right)(t_g - t_a) \tag{3.18}$$

where the heat exchanger effectivity, E_L is given by:

$$E_L = \left(\frac{t_g - t_5}{t_g - t_a}\right) = \left(\frac{\dot{m}_S C_{X_1}}{\dot{m}_W C_{X_4}}\right)\left(\frac{t_3 - t_a}{t_g - t_a}\right) \tag{3.19}$$

The enthalpies at the absorber (inlet and outlet) and enthalpy at the generator outlet are:

$$H_1 = \left(42.81 - 425.92X_1 + 404.67X_1^2\right) + t_a\left(1.01 - 1.23X_1 + 0.48X_1^2\right) \tag{3.20}$$

$$H_5 = \left(42.81 - 425.92X_4 + 404.67X_4^2\right) + t_5\left(1.01 - 1.23X_4 + 0.48X_4^2\right) \tag{3.21}$$

$$H_7 = \left(0.46t_g - 0.043t_c + 572.8\right) \tag{3.22}$$

The main component thermal dynamics and balances are given below:

$$\frac{dQ_C}{dt} = \dot{m}_R(H_7 - H_8) \tag{3.23}$$

$$\frac{dQ_g}{dt} = \dot{m}_W H_5 + \dot{m}_R H_7 - \dot{m}_S H_2 \tag{3.24}$$

$$\frac{dQ_a}{dt} = \dot{m}_W H_6 + \dot{m}_R H_{10} - \dot{m}_S H_1 \tag{3.25}$$

$$\Rightarrow \frac{dQ_c}{dt} + \frac{dQ_a}{dt} = \frac{dQ_g}{dt} + \frac{dQ_e}{dt} \tag{3.26}$$

$$\therefore \frac{dQ_e}{dt} = \frac{dQ_c}{dt} + \frac{dQ_a}{dt} - \frac{dQ_g}{dt} \tag{3.27}$$

The AC performance is gauged using the real and ideal coefficient of performance (COP) along with the relative performance ratio (RPR):

$$COP = \frac{refrigeration\, load}{external\, heat\, input} \tag{3.28}$$

$$COP = \left(\frac{Q_e}{Q_g}\right) = \left(\frac{dQ_e / dt}{dQ_g / dt}\right) = \frac{\left(H_{10} - H_8\right)\left(X_4 - X_1\right)}{X_1 H_5 + (X_4 - X_1) H_7 - X_4 H_1} \tag{3.29}$$

$$COP_i = \frac{T_e\left(T_g - T_a\right)}{T_g\left(T_c - T_e\right)} \tag{3.30}$$

$$RPR = \frac{COP}{COP_i} \tag{3.31}$$

For the GT: the subscripts 3, 4, 5 and 6 denote the components; compressor air inlet, combustor inlet, combustor outlet and turbine exhaust outlet. The compression ratio and work done by the compressor is given as follows:

$$P_4 = rP_3 \tag{3.32}$$

$$\dot{W}_{comp} = \dot{m}_{air} C_{p,air} (T_4 - T_3) \tag{3.33}$$

The combustion chamber temperatures, pressures, heat generated and efficiency is as follows:

$$T_4 = \frac{T_3}{\eta_{comp}}\left[\left(\frac{P_4}{P_3}\right)^{\frac{k-1}{k}} - 1\right] + T_3 \tag{3.34}$$

$$P_5 = P_4 - \Delta P_{comb} \tag{3.35}$$

$$Q_{in} = \dot{m}_{air} C_{p,air}\left(T_5 - T_4\right) \tag{3.36}$$

$$\eta_{cc} = \frac{\dot{Q}_{in} / LHV}{\dot{m}_{fuel}} \tag{3.37}$$

The exhaust discharge temperature and the GT power outputs are as follows:

$$T_6 = T_5 - \eta_{turb} T_4 \left[1 - \left(\frac{1}{P_5 / P_6}\right)^{\frac{k-1}{k}}\right] \tag{3.38}$$

$$\dot{W}_{turb} = \dot{m}_{egy} C_{p,ave}\left(T_5 - T_6\right) \tag{3.39}$$

$$\dot{W}_{net} = \dot{W}_{turb} - \dot{W}_{comp} \tag{3.40}$$

where,

$$\dot{m}_{egy} = \dot{m}_{fuel} + \dot{m}_{air} \tag{3.41}$$

The GT performance indicators are:

$$SFC = \frac{3600\dot{m}_{fuel}}{\dot{W}_{net}} \tag{3.42}$$

$$HR = SFC \cdot LHV \tag{3.43}$$

$$n_{th} = \frac{3600}{SFC \cdot LHV} \tag{3.44}$$

The novel optimization problem is formulated where the constraints were based on the chiller design specifications. Due to the dynamic nature of the model, the objective functions are time-averaged.

Max →Overall Thermal Efficiency, n'_{th}

Min → Heat Rate, HR'

Max →Coefficient of Performance, COP'

subject to the following constraints:

$$\begin{aligned} &32 \le t_a \le 38, 5 \le t_e \le 18, 480 \le t_g \le 1000, 32 \le t_c \le 38, \\ &0 < E_L < 1, 0 \le M_a \le 200, 0 \le M_g \le 200, Q_e \le 2040 \end{aligned} \tag{3.45}$$

As mentioned in Section 2, the PPAs impose an additional constraint to the problem formulation. In this work the minimal amount of power required to be supplied at all time is 129 MW. The PPA considered here allows for a maximal supply of 7.8% above minimal supply power which translates to about 10 MW. Therefore the power supply constraint is obtained as follows:

$$129 \le \dot{W}_{net} \le 139 \tag{3.46}$$

OTIMIZATION STRATEGY: CHAOS-DRIVEN DIFFERENTIAL EVOLUTION (CDDE)

DE is a class of evolutionary meta-heuristic algorithms first introduced in the mid-nineties (Storn and Price, 1995). The incorporation of perturbative methods into evolutionary techniques is the fundamental theme of DE. DE starts by generating a population of a minimum of four individuals denoted, *P*. These individuals are real-coded vectors with some specified size, *N*. The DE algorithm is augmented to enhance its optimization capabilities by the addition of the chaotic component. The chaotic component diversifies the population further enabling it to thoroughly search the objective space.

First, the population of vectors, x^G_i was generated. The consequent steps are similar to the regular DE algorithm where one principal parent, x^p_i and three auxiliary parents x^a_i are randomly selected. Differential mutation is then performed and the mutated vector, V_i is generated. By recombining V_i with x^p_i, the child trial vector, x^{child}_i is created. The obtained x^{child}_i is used as the input to the chaotic map (Flake, 1998). The CDDE approach employed in this work is as follows:

$$N_i(t) = x_i^{child}(t) \tag{3.47}$$

$$R_i(t) = \lambda N_i(t) \tag{3.48}$$

$$N_i(t+1) = R_i(t)N_i(t)\left[1 - N_i(t)\right] \tag{3.49}$$

$$R_i(t+1) = R_i(t) + \lambda' \tag{3.50}$$

where *N(t)* and *R(t)* are variables in the logistic chaotic map, λ' and λ are pre-defined relaxation parameters. Then the logistic mapping is performed until a specific number of iteration is satisfied. The final value at maximum number of iteration of $N(t_{max})$ is incorporated into the child trial vector, x^{child}_i. Hence, the child trial vector, x^{child}_i undergoes another round of mutation by the chaotic map. Survival selection in the next generation is performed via 'knock-out' competition. The fitness function for the child trial vector, x^{child}_i

is evaluated. Thus, another variant of the DE algorithm which is chaotically driven was developed. In this work, this algorithm is called the Chaotic-driven DE (CDDE). The algorithm for the DE and CDDE techniques are given in Algorithm 3.1 and 3.2 respectively:

Algorithm 3.1: Differential Evolution (DE)

Step 1: Set parameters: N, P, CR and F.
Step 2: Random initialization of the population vectors, x^G_i.
Step 3: Random selection of one principal parents, x^p_i
Step 4: Random selection of three auxilary parents, x^a_i
Step 5: Perform differential mutation & generate mutated vector, V_i
Step 6: V_i and x^p_i is recombined to generate child trial vector, x^{child}_i
Step 7: Perform 'knock-out' competition for next generation survival selection
Step 8: *IF* the fitness criterion is satisfied and $t= T_{max}$, halt and print solutions
else proceed to step 3

Algorithm 3.2: Chaos-Driven Differential Evolution (CDDE)

Step 1: Set parameters: N, P, CR and F.
Step 2: Deterministically initialize the population vectors, x^G_i.
Step 3: Iterate chaotic logistic map.
Step 4: *IF* $n > N_{max}$, proceed to next step
else go to Step 3.
Step 5: Randomly select one principal parents, x^p_i
Step 6: Randomly select three auxilary parents, x^a_i
Step 7: Perform differential mutation & generate mutated vector, V_i
Step 8: Recombine V_i with x^p_i to generate child trial vector, x^{child}_i
Step 9: Evaluate fitness of the new x^{child}_i.
Step 10: *IF* the halting conditions are fulfilled halt and print solutions
else proceed to step 2

DOMINANCE MEASURE

As a strictly Pareto-compliant metric, the Hypervolume Indicator (HVI) is often used to measure the quality of solution sets in MO optimization problems (Zitzler *et al.*, 2007). Strictly Pareto-compliant means that if there exist two groups of solutions to some MO problem, then the solution set that dominates the other would give a higher indicator value. The HVI measures the hypervolume of the dominated region of the objective space and can be applied for multi-dimensional scenarios. When using the HVI, a reference point needs to be defined. Relative to this point, the volume of the space of all dominated solutions can be measured. The HVI of a solution set $x_d \in X$ can be defined as follows:

$$HVI(X) = vol\left(\bigcup_{(x_1,\ldots x_d)\in X} [r_1, x_1] \times \ldots \times [r_d, x_d] \right) \tag{3.51}$$

where $r_{1,\ldots,} r_d$ is the reference point and *vol*(.) being the usual Lebesgue measure. In this work the HVI is used to measure the quality of the Pareto frontier obtained by the CDDE and DE techniques. Since the hypervolume values obtained in this work are in the range of 10^{-1}, these values were scaled up to the ranges of 10^3 to get a clearer depiction of the solution quality.

ANALYSES

The triple-objective problem is solved by aggregating the target objectives into a single-objective function – via the weighted-sum approach. The COP and thermal efficiency objectives are in ranges of 0 – 2. The second objective, HR has values in the scale of 10^3. Therefore solving the problem in this form would result in the following:

1. *Biased evaluation of the overall level of dominance* since the value contributed by the HR objective would become heavily wighted as compared to the other two objectives.
2. *Misrepresentation of the aggregate objective function* resulting in misguided fitness tracking of the evolutionary algorithm.

To overcome both these issues the *HR* objective was normalized using the high HR value of 8000 kJ/kWhr:

$$HR' = HR' \,/\, 8000 \tag{3.52}$$

The objective functions are optimized in this work to obtain the chiller design parameters and the amount of GT air inlet cooling required, ΔT. The amount of GT inlet air cooling influences the GT performance as well. Two evolutionary techniques were employed in this research: DE and CDDE in conjunction with the weighted sum approach. The chaotic component was incorporated into the standard DE technique to boost its performance. Additionally the relaxation parameter, λ was varied to increase the degree of chaos in the CDDE. This was done to further diversify the generated solutions to obtain more dominant Pareto frontiers. The relaxation is reduced by increasing the parameter λ which increases the degree of chaos. To determine the dominance levels of the Pareto frontiers, the nadir point (0.1, 1, 0.1) was employed.

The algorithms implemented in this work were developed using the C++ programming language on a personal computer with an Intel® Core ™ i5 processor running at 3.2 GHz. The entire Pareto frontier was constructed using compromised individual solutions obtained at different scalarizations. Due to stochastic nature of the algorithm, the best individual solution was taken after five independent runs. The optimization carried out in this work was on a dynamic system for a 24 hour cycle. Therefore the objective values obtained here are time-averaged. The individual solutions obtained are graded by taking the best, median and worst solution across the frontier. The graded solutions and the efficient Pareto frontier generated by the DE approach are given in Table 1 and Figure 2.

Tables 2-4 present the graded individual solutions produced by the CDDE obtained at various degrees of chaos while Figure 3 illustrates the corresponding Pareto frontiers.

In terms of the best individual solution dominance, the CDDE technique at $\lambda = 0.5$ ranks highest as compared to the conventional DE approach by 19.275% (Tables 3.1 and 3.3). In Tables 3 and 4, it can be seen that the dominance level of the best individual solutions of the CDDE techniques at $\lambda = 0.5$ and 0.9 are very close (with a difference of less than 1%) and significantly higher than the CDDE technique at $\lambda = 0.01$. Therefore higher degrees of chaos in the CDDE technique improve the level of individual

Table 1. The graded individual solutions obtained using the DE technique

Description		Best	Median	Worst
Objective Functions	n'_{th}	0.3358	0.3278	0.3256
	HR'	0.7272	0.7449	0.75
	COP'	1.8822	1.883	1.8959
Decision Variable	t_a	37.131	37.2141	37.1113
	t_e	8.57596	8.5671	8.26275
	t_g	481.576	481.567	481.263
	t_c	36.424	36.4329	36.7373
	E_L	0.9984	0.9956	0.8993
	M_a	107.069	107.812	103.74
	M_g	107.069	107.812	103.74
	ΔT	7.0693	2.4704	1.1827
Output Parameters	Q_E	722.919	192.646	96.2117
	W_{net}	134.303	131.126	130.237
Metric	HVI	114.6052	103.6377	101.3062

Figure 2. Pareto frontier generated by the DE technique

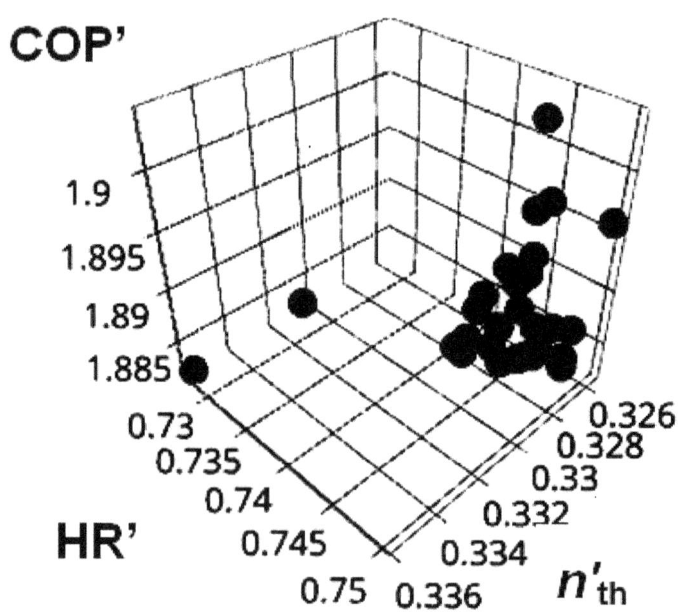

Table 2. The graded individual solutions obtained using the CDDE technique at λ = 0.01

Description		**λ = 0.01**		
		Best	**Median**	**Worst**
Objective Functions	n'_{th}	0.3384	0.33	0.3242
	HR'	0.722	0.740	0.753
	COP'	1.936	1.939	1.958
Decision Variable	t_a	35.343	35.465	37.877
	t_e	7.696	7.439	7.031
	t_g	486.957	484.394	480.307
	t_c	37.304	37.561	37.969
	E_L	0.570	0.544	0.503
	M_a	104.300	101.859	100.184
	M_g	104.300	101.859	100.184
	ΔT	8.580	3.710	0.368
Output Parameters	Q_E	482.601	459.290	27.114
	W_{net}	135.347	131.982	129.674
Metric	HVI	121.7962	109.9217	102.7748

solution dominance. The best individual solution (see bold-faced column in Table 3.3) achieved by the CDDE at $\lambda = 0.5$ varies very slightly with the CDDE at $\lambda = 0.9$ – where it maximizes n'_{th} and minimizes the HR' by 0.001. It is fair to state that for this scenario, when $\lambda = 0.5$, the optimal region in the search space has been reached. Further increasing the chaos, $\lambda > 0.5$ only improves the solution to a very small degree. The overall dominance of the entire Pareto frontier produced the evolutionary techniques produced in this work is depicted in Figure 3.4:

In Figure 4 the frontier dominance of the CDDE approaches are higher than that of the conventional DE technique. Among the CDDE techniques it can be observed that the higher the level of chaos in the algorithm, the higher the degree of the dominance of the entire frontier. The frontier produced by CDDE with $\lambda = 0.9$ (most dominant frontier) outweighs that of the conventional DE technique by 9.9313%. The frontier produced by CDDE with $\lambda = 0.9$ outweighs that produced by CDDE with $\lambda = 0.5$ and $\lambda = 0.01$ by 1.0561% and 3.2026% respectively. The overall frontier dominance trend follows the trend observed in the dominance of the individual solution points produced

Table 3. The graded individual solutions obtained using the CDDE technique at λ = 0.5

Description		λ = 0.5		
		Best	**Median**	**Worst**
Objective Functions	n'_{th}	**0.3485**	0.3299	0.3246
	HR'	**0.701**	0.740	0.752
	COP'	**1.937**	1.936	1.913
Decision Variable	t_a	**35.949**	35.127	31.727
	t_e	**7.930**	7.470	7.656
	t_g	**489.298**	484.704	486.564
	t_c	**37.070**	37.530	37.344
	E_L	**0.593**	0.547	0.566
	M_a	**107.247**	101.831	100.291
	M_g	**107.247**	101.831	100.291
	ΔT	**14.461**	3.654	0.580
Output Parameters	Q_E	**389.379**	522.700	1082.500
	W_{net}	**139.409**	131.944	129.820
Metric	*HVI*	**136.6954**	109.6404	100.8299

by the techniques employed in this work. The frontier produced by the DE technique contains many solutions which are scattered in non-optimal (or sub-optimal) regions of the objective space (Figure 3.2). As for the CDDE approach, it could be seen in Figure 3 that as the chaotic level increases, the solutions constructing the Pareto frontier gravitate toward optimal region in the objective space. When many of the individual solution have these characteristics, the overall frontier dominance significantly increases as a whole – which is the case for solutions produced by the CDDE at $\lambda = 0.9$. The computational effort taken by each of the techniques employed in this work is assessed by analyzing their execution time. The execution time for all the techniques is depicted in Figure 5:

The CDDE approach takes the shortest time to construct the entire Pareto frontier as compared to DE. The CDDE approach at $\lambda = 0.9$ generates the optimal Pareto frontier at approximately 3 times faster than the conventional DE technique. It should be noted that incorporating the chaotic component into the conventional DE approach to create the CDDE increases the algorithmic complexity of the technique. Accounting for increased algorithmic complexity,

Table 4. The Graded individual solutions obtained using the CDDE technique at λ = 0.9

Description		λ = 0.9		
		Best	**Median**	**Worst**
Objective Functions	n'_{th}	0.3478	0.3286	0.3249
	HR'	0.702	0.743	0.751
	COP'	1.940	1.950	1.917
Decision Variable	t_a	36.326	36.916	32.266
	t_e	7.870	7.254	7.614
	t_g	488.704	482.542	486.135
	t_c	37.130	37.746	37.387
	E_L	0.587	0.525	0.561
	M_a	107.029	101.458	100.401
	M_g	107.029	101.458	100.401
	ΔT	14.025	2.908	0.800
Output Parameters	Q_E	1161.140	208.804	1006.490
	W_{net}	139.108	131.428	129.972
Metric	*HVI*	135.8047	108.5819	101.5729

the CDDE techniques still achieves the optimal solutions faster than the less complex conventional DE algorithm. Therefore the effectiveness of the chaotic component during search outweighs the setbacks from additional computational effort due to algorithmic complexity. All evolutionary approaches employed in this work performed stable computations without any stagnation in the objective space during program executions. All solution points used for constructing the Pareto frontier were feasible and none of the constraints presented in the problem was violated. The salient feature of the chaotic component is that it enhances the evolutionary technique to increase the rigor of its search in an efficient manner. Additionally, the diversification capability of the evolutionary algorithm could be controlled in a deterministic manner via the chaos relaxation parameter, λ.

The PPA puts a cap on the power output as shown in equation (3.40). Therefore optimization of the GT falls back on the HR which lowers the fuel consumption. The mass of fuel consumed by the GT per kWh of power generated is attained by calculating the specific fuel consumption (SFC in kg/kWh). The SFC produced by each of the algorithms is shown in Figure 6.

Figure 3. Pareto frontiers generated by the CDDE technique at varying levels of chaos

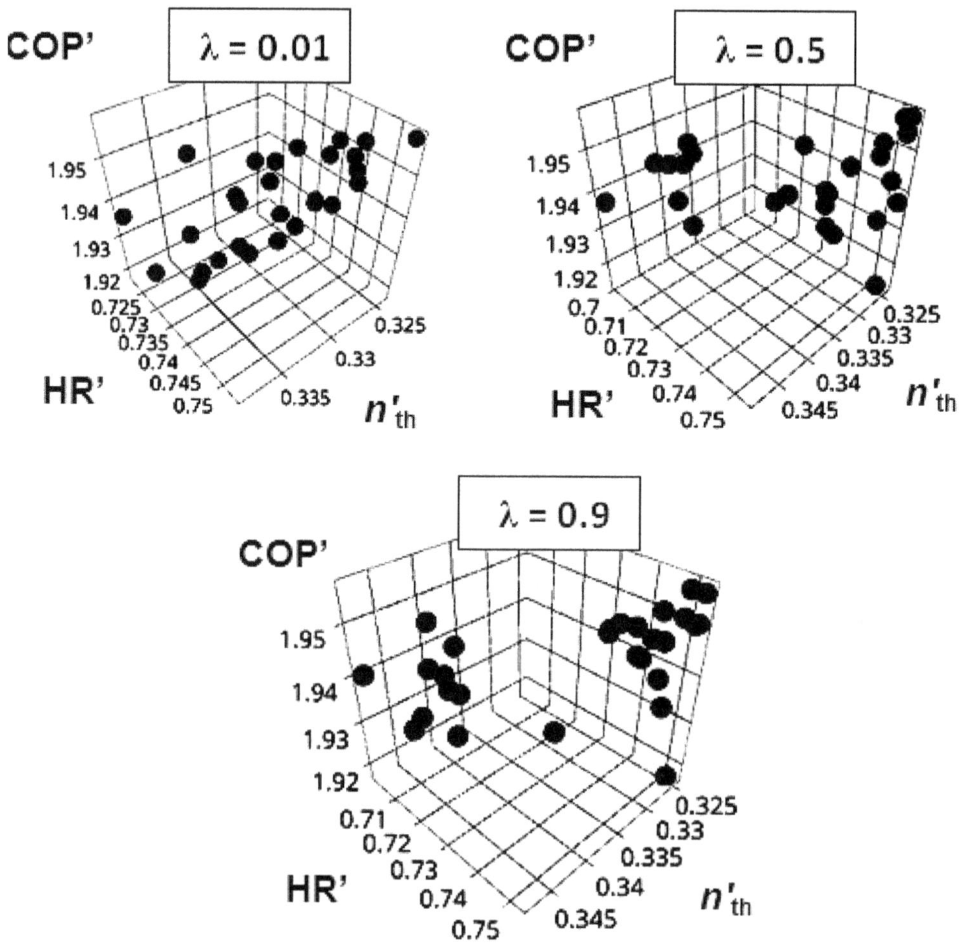

The time-averaged SFC values of the best solutions produced by the CEDE techniques at $\lambda = 0.5$ and $\lambda = 0.9$ are very close. The SFC produced by the CEDE at $\lambda = 0.5$ outweighs the DE technique by 3.6636%.

Using the CDDE approach, a new optimal design and operation of the combined GT-AC system is reached. Besides single point solutions, an entire optimal frontier of compromised solutions was generated by using evolutionary strategies in tandem with the weighted sum approach. The chaotic enhancement was observed to significantly boost the performance of the DE approach in two respects; (1) minimizing the computational effort and (2) maximizing the efficiency of reaching the optima. Interestingly the most optimal individual solution (see bold-faced column in Table 3.3) consists

Figure 4. Overall dominance levels of the entire Pareto frontiers obtained by the evolutionary techniques

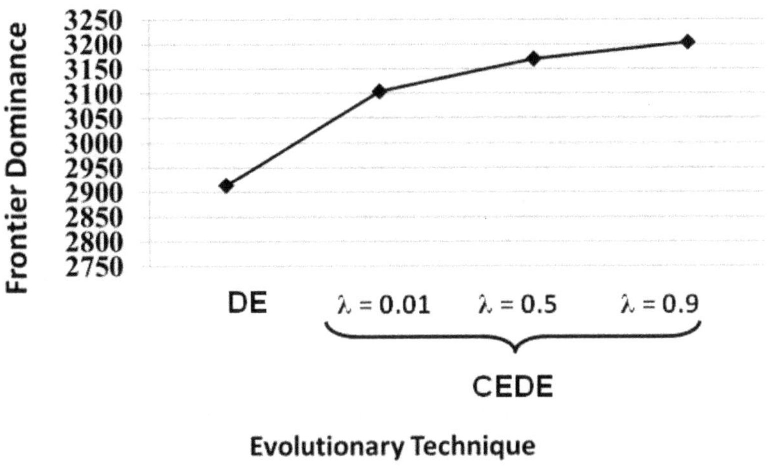

cooling the inlet air to the turbine with a temperature difference of about 14 °C while optimizing all other objective functions. The fuel consumption of the GT was observed to be significantly improved as a result from the optimization (via the implementation of the CEDE approach).

Figure 5. Computational time taken by the evolutionary techniques

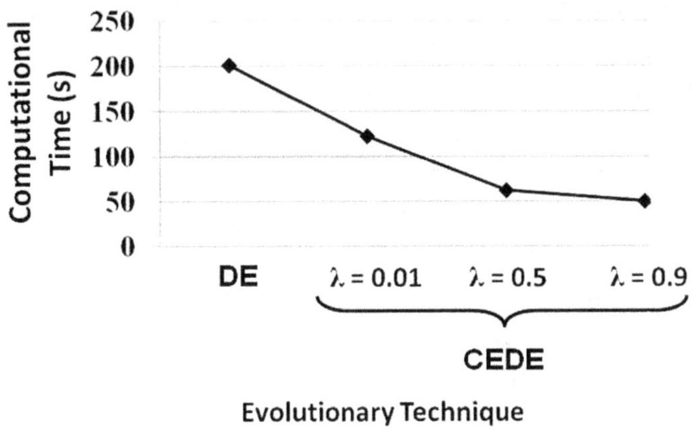

Figure 6. Specific fuel consumption (SFC) produced by the best individual solutions

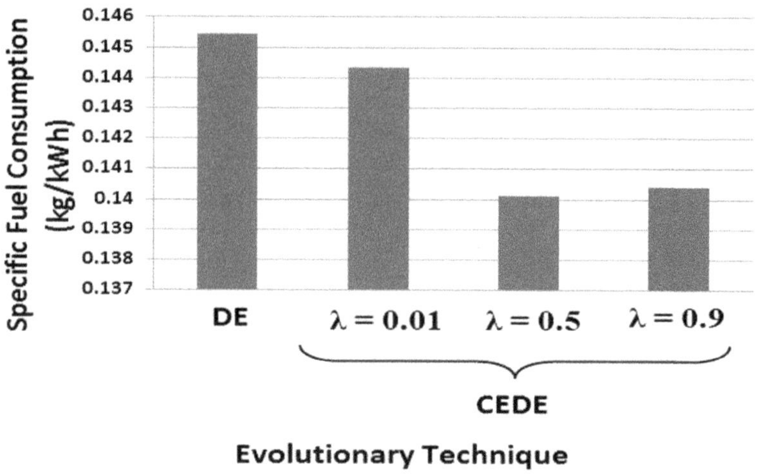

PART 2: OPTIMIZATION OF THE THREE-OBJECTIVE GT-AC SYSTEM UNDER UNCERTAINTIES IN WEATHER CONDITIONS

Overview

This part presents the novel modeling of a real-world combined GT-AC multicriteria optimization problem under uncertainty. Similar to the previous parts of this chapter, ambient air conditions directly influence the turbine inlet air temperature. To account for uncertainties in the ambient air temperature, the combined GT-AC system was modeled using a type-2 fuzzy programming approach. This multicriteria fuzzy optimization problem was solved using the chaotic-driven differential evolution strategy (CDDE) (see Section 3.4). Energy systems that rely on their surroundings are often difficult to design if the surroundings contain irregularities or uncertainties. In GT-AC systems, the ambient temperature is a weather-dependent variable. Using meteorological data, the type-2 FL is employed to model and incorporate ambient temperature considerations into the problem formulation. The meteorological information for Sepang – Kuala Lumpur International Airport AB (KLIA) at Malaysia was retrieved from the weather database (WE, 2016). The daily average ambient temperature (in °C) was obtained for the month of January in the year 2016. Table 5 provides the monthly ambient temperature.

The primary membership function, $\eta_F(x)$ was employed to model the weekly data while the secondary membership function, $\mu_F(y, \eta_F(x))$ was used to model the overall monthly data. The S-curve function is employed as the primary and secondary membership functions ($\eta_F(x)$ and $\mu_F(y, \eta_F(x))$). Therefore type-2 fuzzification is performed on the ambient temperature using the S-curve membership function. This is carried out by determining the average, maximum and minimum values of ambient temperature from the meteorological data. The same averaged dynamic model which is given in equations 3.1 - 3.46 was employed.

Table 5. Daily average ambient temperature (°C) obtained for January 2016

DAY	**1**	**2**	**3**	**4**	**5**	**6**	**7**
Mean	26	28	28	28	29	30	30
Max	30	33	32	33	34	34	34
Min	23	23	24	24	24	25	25
DAY	**8**	**9**	**10**	**11**	**12**	**13**	**14**
Mean	30	29	29	29	29	29	29
Max	34	33	34	34	34	34	34
Min	25	25	24	24	24	24	24
DAY	**15**	**16**	**17**	**18**	**19**	**20**	**21**
Mean	28	28	28	29	28	30	29
Max	32	32	31	33	33	36	34
Min	24	24	24	25	24	24	24
DAY	**22**	**23**	**24**	**25**	**26**	**27**	**28**
Mean	29	28	29	30	30	29	30
Max	34	34	34	35	33	33	34
Min	24	23	24	25	26	25	25
DAY	**29**	**30**					
Mean	28	29					
Max	33	34					
Min	24	24					

COMPUTATIONAL RESULTS

The fuzzy multicriteria problem was solved via the weighted-sum approach. The objective functions are optimized to obtain the chiller design parameters and the amount of GT air inlet cooling required, ΔT - while accounting for uncertainties in the ambient temperature data in the optimization model. The CEDE technique was employed in this work was incorporated with the chaotic component to boost its optimization performance. The relaxation parameter, λ was varied to increase the degree of chaos in the CEDE. After rigorous testing, it was identified that the most optimal frontier is obtained when the CEDE is tuned with maximal chaotic capability (at $\lambda = 0.9$). The nadir point (0.1, 1, 0.1) was employed as a reference value for the HVI. The algorithms implemented in this work were developed using the C++ programming language on a personal computer with an Intel® Core ™ i5 processor running at 3.2 GHz.

The entire Pareto frontier was constructed using compromised individual solutions obtained at different scalarizations. Due to the stochastic nature of the algorithm, the best individual solution was taken after five independent runs. The optimization carried out in this work was on a dynamic system for a 24-hour cycle. Therefore the objective values obtained here are time-averaged. The fuzzified ambient temperature with respect to various membership grades is presented in Figure 7.

The membership grade $\mu_F(y, \eta_F(x))$ is a tensor since it takes vector values of $\eta_F(x)$ as inputs. To reduce its dimensions, $\mu_F(y, \eta_F(x))$ values are averaged back into a vector form as depicted in Figure 7. The ranges for the membership grades are as follows:

$$\mu_{F,S} = [0.00546\ 0.19579]$$

$$\eta_{F,S} = [0.06648, 0.17169] \tag{3.53}$$

Using a type-2 fuzzy formulation to account for the uncertainties in the ambient temperature, the CEDE technique was applied to the combined MO GT-AC system. The optimal values for the overall thermal efficiency, $\overline{\overline{n}}'_{th}$ is given in Figure 8.

Based on Figure 8, the variation in the thermal efficiency of the GT at all membership grades is 6.9842% (corresponding to 0.04365 differences in

Figure 7. Fuzzy ambient temperature versus membership grades

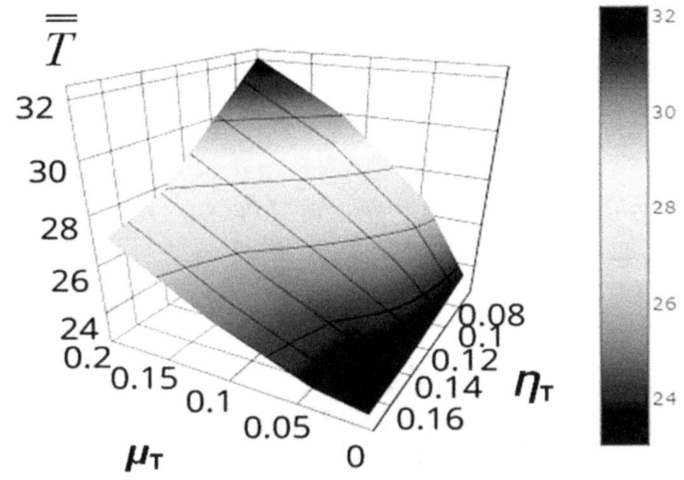

Figure 8. Overall thermal efficiency versus membership grades

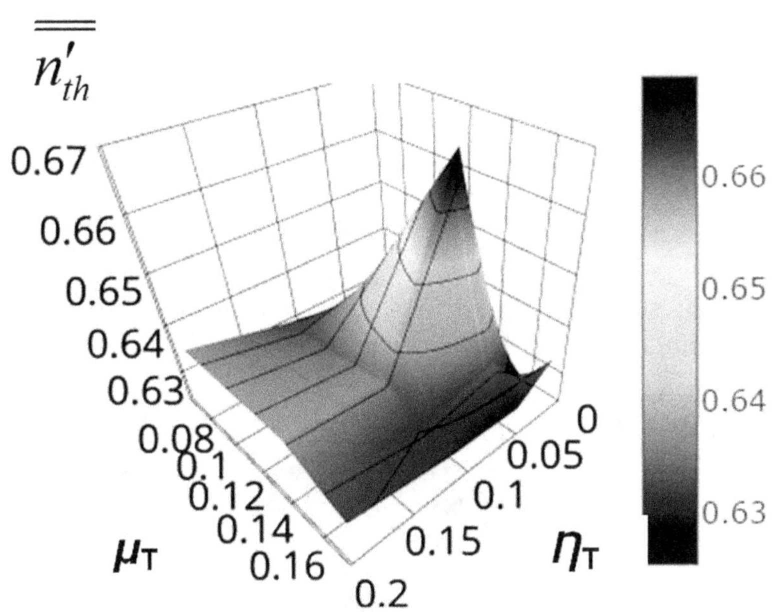

thermal efficiency). A maximum thermal efficiency of 0.6686 was obtained at $\mu_{F,T} = 0.04298$ and $\eta_{F,T} = 0.13811$. The minimal thermal efficiency at $\mu_{F,T} = 0.00547$ and $\eta_{F,T} = 0.15725$ was 0.62499. Figure 9 and 10 depicts the normalized HR and the COP respectively relative to the primary and secondary membership values.

The maximal variation in terms of the *HR* relative to the membership grades is 376.06 kJ/kWhr. The maximum normalized *HR* is 0.72003 (5760.27

Figure 9. Normalized HR versus membership grades

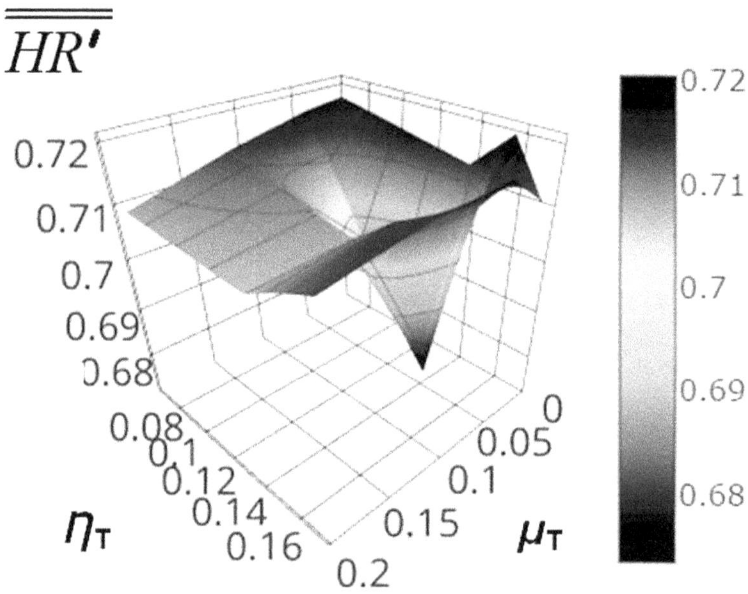

Figure 10. Overall thermal efficiency versus membership grades

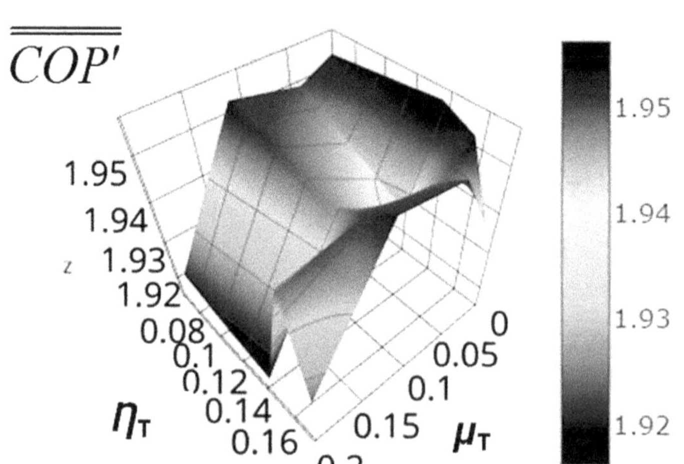

kJ/kWhr) at $\eta_{F,T} = 0.15725$ and $\mu_{F,T} = 0.00547$. The minimal normalized *HR* is 0.67302 (5384.22 kJ/kWhr) at $\eta_{F,T} = 0.13811$ and $\mu_{F,T} = 0.04298$. On the other hand, the *COP* varies by 2.08991% (0.04). It can observed that the uncertainties in temperature variation significantly impacts *HR* and thermal efficiency of the GT while affecting the *COP* of the AC to a lesser degree (< 5%). From these findings it could be said that uncertainty in ambient temperatures highly influences the optimization on GT operations as compared to the design of the AC. The temperature data considered here is only for over a range in a single month. When projected further to an annual scale, the minor variations in the COP may translate to significant cost savings. Besides, this work focuses on a single unit combined GT-AC system; these variations in *COP*, thermal efficiency and *HR* may compound to produce a greater impact when considering multiple GT-AC systems. The solutions obtained for various membership grades are measured using the HVI. The membership grade with the highest level of dominance in the objective space is $\mu_{F,T} = 0.04298$ and $\eta_{F,T} = 0.13811$. The Pareto frontier obtained at various scalar values using the weighted-sum approach is given in Figure 11. The ranked solutions (best, median and worst) of the Pareto frontier are given in Table 6.

The overall frontier dominance trend follows the trend observed in the dominance of the individual solution points produced by the techniques employed in this work. It can be seen that in Figure 11, the solutions are diversely spread throughout the objective space. The computational time taken for the complete construction of the Pareto frontier for each membership grade is approximately 50 – 60 seconds. The CEDE approach in this work performed stable computations exploring the objective space smoothly during program executions. All solution points used for constructing the Pareto frontier were feasible and no constraints were broken. The chaotic component was seen to boost the performance of the conventional DE technique enabling it to perform a more thorough search - by controlling the relaxation parameter, λ.

The PPA considered in this work limits the power output of the GT. Further optimization results in the minimization of the *HR* which increases the efficiency of the GT by lowering the fuel consumption. During the optimization of the GT (while considering the ambient temperature), the *HR* was observed to significantly vary (see Figure 9). The mass of fuel consumed by the GT per kWh of power generated is attained by calculating the specific fuel consumption - *SFC* (in kg/kWh). The *SFC* could be determined from

Figure 11. Scatter plot depicting the Pareto frontier at the membership values of $\mu_{F,T} = 0.04298$ *and* $\eta_{F,T} = 0.13811$.

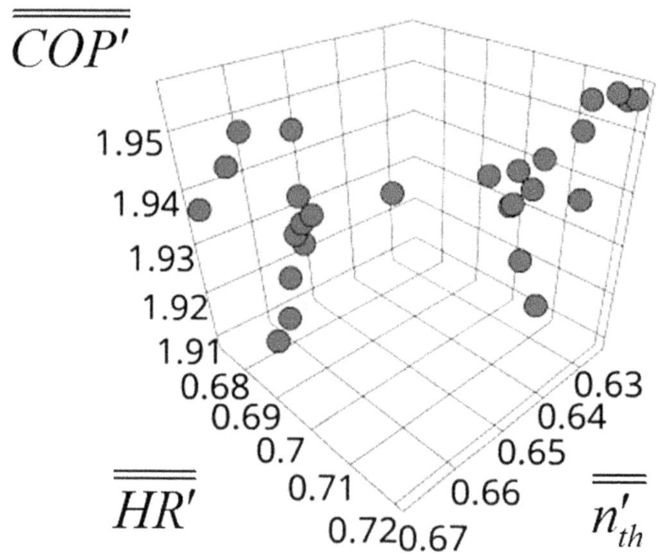

Table 6. Ranked solutions for Pareto frontier obtained at Pareto frontier at the membership values of $\mu_{F,T} = 0.04298$ *and* $\eta_{F,T} = 0.13811$.

Description		Best	Median	Worst
Objective Functions	n_{th}	0.668635	0.647951	0.624774
	HR	0.673027	0.694512	0.720277
	COP	1.93662	1.93949	1.95522
Decision Variable	t_a	35.8297	35.8152	37.5478
	t_e	7.90137	7.5902	7.06663
	t_g	489.014	485.902	480.666
	t_c	37.0986	37.4098	37.9334
	E_L	0.590137	0.55902	0.506663
	M_a	106.843	103.717	100.214
	M_g	106.843	103.717	100.214
	ΔT	13.6545	7.41669	0.427082
Parameters	Q_E	379.245	395.578	88.7511
	$\dot{W}_{net}$	133.727	129.59	124.955
Metric	HVI	341.4796	307.9167	272.3303

the *HR*. In Figure 9, the maximum *HR* of 5760.27 kJ/kWhr corresponds to 0.144 kg/kWh while the minimum *HR* of 5384.22 kJ/kWh is about 0.1346 kg/kWh. Therefore the *SFC* has a reduction of 6.528% from the maximum at different membership grades during optimization – while considering uncertainties in the ambient temperature.

The type-2 fuzzy framework was implemented to optimize the design and operations of the combined GT-AC system. The ambient temperature data contained two moments of uncertainty making it suitable to be captured using type-2 fuzzy inference. Using the CEDE evolutionary technique the fuzzy formulation was effectively solved with the aid of the weighted sum approach. The solutions obtained maximized the GT thermal efficiency and *HR* while minimizing the *COP* of the AC. Due to the GT power output constraint imposed by the PPA, the optimization directly impacted the *HR* – resulting in lower fuel consumption (*SFC*)

INSIGHTS

A new optimal design for the combined GT-AC system was achieved using the CEDE approach in Part 1. To improve the optimization techniques, the chaotic component could be employed in other metaheuristics as an effective tool to boost the optimization. In this part the dynamic model was optimized by considering a fixed time-averaged ambient temperature distribution. Forecasting and regression modeling methods such as robust conic multivariate adaptive regression splines (RCMARS) could also be employed in tandem with the metaheuristic (Özmen *et al.*, 2013; Özmen *et al.*, 2014). In addition, other types of metaheuristic could be tested on this problem – e.g. swarm algorithms; Particle Swarm Optimization (Kennedy & Eberhart, 1995), Bacteria Foraging (Liu *et al.*, 2002) and Cuckoo Search (Yildiz, 2013).

REFERENCES

Ahmadi, P., Dincer, I., & Rosen, M. A. (2013). Thermodynamic modeling and multi-objective evolutionary-based optimization of a new multigeneration energy system. *Energy Conversion and Management*, *76*, 282–300. doi:10.1016/j.enconman.2013.07.049

Beyer, H. G. (2013). *The theory of evolution strategies*. Springer Science & Business Media.

Bierdel, B. H. T. (2013). *Life Cycle Value for Combined Cycle Power Plants*. Siemens AG.

Brito, T. G., Paiva, A. P., Ferreira, J. R., Gomes, J. H. F., & Balestrassi, P. P. (2014). A normal boundary intersection approach to multiresponse robust optimization of the surface roughness in end milling process with combined arrays. *Precision Engineering*, *38*(3), 628–638. doi:10.1016/j.precisioneng.2014.02.013

Cory, K., Canavan, B., & Koenig, R. (2009). *Power Purchase Agreement Checklist for State and Local Governments (No. NREL/FS-6A2-46668). National Renewable Energy Laboratory*. Golden, CO: NREL. doi:10.2172/967194

Deb, K., Pratap, A., Agarwal, S., & Meyarivan, T. (2002). A Fast and Elitist Multiobjective Genetic Algorithm: NSGA-II. *IEEE Transactions on Evolutionary Computation*, *6*(2), 182–197. doi:10.1109/4235.996017

Debbarma, S., Saikia, L. C., & Sinha, N. (2014). Robust two-degree-of-freedom controller for automatic generation control of multi-area system. *International Journal of Electrical Power & Energy Systems*, *63*, 878–886. doi:10.1016/j.ijepes.2014.06.053

Flake, G. W. (1998). *The computational beauty of nature: Computer explorations of fractals, chaos, complex systems, and adaptation*. Cambridge, MA: MIT Press.

Ganesan, T., Aris, M. S., & Elamvazuthi, I. (2018). Multiobjective Strategy for an Industrial Gas Turbine: Absorption Chiller System. In Handbook of Research on Emergent Applications of Optimization Algorithms (pp. 531-556). IGI Global.

Ganesan, T., Aris, M. S., Elamvazuthi, I., & Tageldeen, M. K. (2016). *Type–2 Fuzzy Programming for Optimizing the Heat Rate of an Industrial Gas Turbine via Absorption Chiller Technology*. Retrieved from https://www.semanticscholar.org/paper/Type-%E2%80%93-2-Fuzzy-Programming-for-Optimizing-the-Heat-Elamvazuthi-Tageldeen/ab3e33202cc873e823bedb87671f2bda5b0400e5

Ganesan, T., & Elamvazuthi, I. (2017). A multi-objective approach for resilience-based plant design optimization. *Quality Engineering*, *29*(4), 656–671. doi:10.1080/08982112.2016.1255331

Ganesan, T., Elamvazuthi, I., & Vasant, P. (2015). Multiobjective design optimization of a nano-CMOS voltage-controlled oscillator using game theoretic-differential evolution. *Applied Soft Computing*, *32*, 293–299. doi:10.1016/j.asoc.2015.03.016

Ganesan, T., Elamvazuthi, I., Zilati Ku Shaari, K., & Vasant, P. (2014). Hopfield differential evolution for multi-objective optimization of a cement-bonded sand mould system. *International Journal of Management Science and Engineering Management*, *9*(1), 40–47. doi:10.1080/17509653.2013.858416

García-Revillo, F. J., Jimenez-Octavio, J. R., Sanchez-Rebollo, C., & Cantizano, A. (2014). Efficient multi-objective optimization for gas turbine discs. In *Design and Computation of Modern Engineering Materials* (pp. 227–255). Springer International Publishing. doi:10.1007/978-3-319-07383-5_17

Grefenstette, J. J. (2013). Genetic Algorithms and Their Applications. In *Proceedings of the Second International Conference on Genetic Algorithms*. Psychology Press. 10.4324/9780203761595

Javadi, M. S., & Javadinasab, A. (2011). Power Purchasing Agreements in Modern Power System. *The Journal of American Science*, *7*(6), 164–169.

Jiang, S., Zhang, J., Ong, Y. S., Zhang, A. N., & Tan, P. S. (2015). A simple and fast hypervolume indicator-based multiobjective evolutionary algorithm. *IEEE Transactions on Cybernetics*, *45*(10), 2202–2213. doi:10.1109/TCYB.2014.2367526 PMID:25474815

Kennedy, J., & Eberhart, R. (1995). Particle Swarm Optimization. *IEEE Proceedings of the International Conference on Neural Networks,* 1942-1948.

Khorasani Nejad, E., Hajabdollahi, F., Hajabdollahi, Z., & Hajabdollahi, H. (2013). Thermo-economic Optimization of Gas Turbine Power Plant with Details in Intercooler. *Heat Transfer—Asian Research, 42*(8), 704-723.

Kohlenbach, P., & Ziegler, F. (2008). A dynamic simulation model for transient absorption chiller performance. Part I: The model. *International Journal of Refrigeration*, *31*(2), 217–225.

Labrador, D. (2015). *Rocky Mountain Institute Outlet*. Available: http://blog.rmi.org/blog_2015_04_15_corporate_power_purchasing_is_poised_to_be_the_next_big_thing_in_renewable_energy

Liu, G., Wu, G., Zheng, T., & Ling, Q. (2011, July). Integrating preference based weighted sum into evolutionary multi-objective optimization. In *Natural Computation (ICNC), 2011 Seventh International Conference on* (Vol. 3, pp. 1251-1255). IEEE. 10.1109/ICNC.2011.6022362

Liu, Y., Passino, K. M., & Simaan, M. A. (2002). Biomimicry of social foraging bacteria for distributed optimization: Models, principles, and emergent behaviors. *Journal of Optimization Theory and Applications*, *115*(3), 603–628. doi:10.1023/A:1021207331209

Marler, R. T., & Arora, J. S. (2010). The weighted sum method for multi-objective optimization: New insights. *Structural and Multidisciplinary Optimization*, *41*(6), 853–862. doi:10.100700158-009-0460-7

Mohanty, B., Panda, S., & Hota, P. K. (2014). Differential evolution algorithm based automatic generation control for interconnected power systems with non-linearity. *Alexandria Engineering Journal*, *53*(3), 537–552. doi:10.1016/j.aej.2014.06.006

Naidu, K., Mokhlis, H., & Bakar, A. H. A. (2014). Multiobjective optimization using weighted sum artificial bee colony algorithm for load frequency control. *International Journal of Electrical Power & Energy Systems*, *55*, 657–667. doi:10.1016/j.ijepes.2013.10.022

Nasiruddin, I., Bhatti, T. S., & Hakimuddin, N. (2015). Automatic generation control in an interconnected power system incorporating diverse source power plants using bacteria foraging optimization technique. *Electric Power Components and Systems*, *43*(2), 189–199. doi:10.1080/15325008.2014.975871

Negara, P. L. (2013). *Power In Indonesia Investment and Taxation Guide*. Academic Press.

Ning, B., & Bloomfield, P. (2017). *Bayesian inference for generalized extreme value distribution with Gaussian copula dependence*. arXiv preprint arXiv:1703.00968

Oliveira, L. S. D., & Saramago, S. F. (2010). Multiobjective optimization techniques applied to engineering problems. *Journal of the Brazilian Society of Mechanical Sciences and Engineering*, *32*(1), 94–105. doi:10.1590/S1678-58782010000100012

Özmen, A., & Weber, G. W. (2014). RMARS: Robustification of multivariate adaptive regression spline under polyhedral uncertainty. *Journal of Computational and Applied Mathematics*, *259*, 914–924. doi:10.1016/j.cam.2013.09.055

Özmen, A., Weber, G. W., Çavuşoğlu, Z., & Defterli, Ö. (2013). The new robust conic GPLM method with an application to finance: Prediction of credit default. *Journal of Global Optimization*, *56*(2), 233–249. doi:10.100710898-012-9902-7

Plucar, J., Grunt, O., & Zelinka, I. (2017, July). Case study: Optimizing fault model input parameters using bio-inspired algorithms. In AIP Conference Proceedings (Vol. 1863, No. 1, p. 070027). AIP Publishing.

Price, K., Storn, R. M., & Lampinen, J. A. (2006). *Differential evolution: a practical approach to global optimization*. Springer Science & Business Media.

Rao, S. S., & Rao, S. S. (2009). *Engineering optimization: theory and practice*. John Wiley & Sons. doi:10.1002/9780470549124

Rout, U. K., Sahu, R. K., & Panda, S. (2013). Design and analysis of differential evolution algorithm based automatic generation control for interconnected power system. *Ain Shams Engineering Journal*, *4*(3), 409–421. doi:10.1016/j.asej.2012.10.010

Shao, L., Liu, L., & Li, X. (2014). Feature learning for image classification via multiobjective genetic programming. *IEEE Transactions on Neural Networks and Learning Systems*, *25*(7), 1359–1371. doi:10.1109/TNNLS.2013.2293418

Singh, R. P., Mukherjee, V., & Ghoshal, S. P. (2015). Optimal reactive power dispatch by particle swarm optimization with an aging leader and challengers. *Applied Soft Computing*, *29*, 298–309. doi:10.1016/j.asoc.2015.01.006

Storn, R., & Price, K. V. (n.d.). *Differential evolution – a simple and efficient adaptive scheme for global optimization over continuous spaces*. ICSI, Technical Report TR-95-012.

Sun, D. D., Yang, C., & Zeng, F. (2014). Study on Gas Turbine-Based CCHP System with Multi-Objective Evaluation Index. *Advanced Materials Research*, *860*, 1366–1369.

Wu, O. Q., & Babich, V. (2012). Unit-contingent power purchase agreement and asymmetric information about plant outage. *Manufacturing & Service Operations Management: M & SOM*, *14*(2), 245–261. doi:10.1287/msom.1110.0362

Yazdi, B. A., Yazdi, B. A., Ehyaei, M. A., & Ahmadi, A. (2015). Optimization of Micro Combined Heat And Power Gas Turbine By Genetic Algorithm. *Thermal Science*, *19*(1), 207–218. doi:10.2298/TSCI121218141Y

Yildiz, A. R. (2013). Cuckoo search algorithm for the selection of optimal machining parameters in milling operations. *International Journal of Advanced Manufacturing Technology*, *64*(1-4), 55–61. doi:10.100700170-012-4013-7

Zitzler, E., Brockhoff, D., & Thiele, L. (2007, March). The hypervolume indicator revisited: On the design of Pareto-compliant indicators via weighted integration. In *International Conference on Evolutionary Multi-Criterion Optimization* (pp. 862-876). Springer Berlin Heidelberg. 10.1007/978-3-540-70928-2_64

APPENDIX

Nomenclature and Abbreviations for the GT-AC System

Subscripts

1: *stream from absorber to pump*
2: *stream from pump to solution HEX*
3: *stream from HEX to generator*
4: *stream from generator to HEX*
5: *stream from HEX to valve*
6: *stream from valve to absorber*
7: *stream from generator to condenser*
8: *stream from condenser to expansion valve*
9: *stream from expansion valve to evaporator*
1: *stream from evaporator to absorber*

Absorption Chiller

X_s: *strong solution concentration* (kg LiBr/ kg solution)
X_w: *weak solution concentration* (kg LiBr/ kg solution)
t_a: *absorber temperature* (°C)
t_e: *evaporator temperature* (°C)
t_g: *generator temperature* (°C)
t_c: *condenser temperature* (°C)
P_e: *evaporator pressure* (mm Hg)
P_c: *condenser pressure* (mm Hg)
H_i : *enthalpy of pure water* (kcal/kg)
Q_E: *transient refrigeration load* (kcal/hr)
$\dot{m}_R$: *refrigerant mass flowrate* (kg/hr)
$\dot{m}_W$: *weak LiBr solution mass flowrate* (kg/hr)
$\dot{m}_S$: *strong LiBr solution mass flowrate* (kg/hr)
E_L : *HEX effectivity*
C_{X_1} : *specific heat of the strong solution* (kcal/kg °C)
C_{X_4} : *specific heat of the weak solution* (kcal/kg °C)
Q_e : *heat input to the evaporator* (kcal)

Q_g : *heat input to the generator* (kcal)
Q_c : *heat output from the evaporator* (kcal)
Q_a : *heat output from the absorber* (kcal)
dQ_e / dt : *heat rate - input to the evaporator* (kcal/hr)
dQ_g / dt : *heat rate - input to the generator* (kcal/hr)
dQ_c / dt : *heat rate - output to the evaporator* (kcal/hr)
dQ_e / dt : *heat rate - input to the absorber* (kcal/hr)
COP: *coefficient of performance*
COP_i: *ideal/max coefficient of performance*
RPR: *relative performance ratio*
$\dot{m}_{a,t}(i)$: *outlet mass flowrate at the absorber tube* (kg/hr)
$\dot{m}_{g,t}(i)$: *outlet mass flowrate at the generator tube*(kg/hr)
$M_a(i)$: *mass stored at the absorber sump (shell side) tube* (kg)
$M_g(i)$: *mass stored at the generator sump (shell side) tube*(kg)
Δt : *simulation time-interval* (hr)
COP': *time-averaged coefficient of performance*
ΔT : *temperature differential for cooling* (hr)

Gas Turbine

p_i: *component pressures (kPa)*
r: compression pressure ratio
η_{comp} : *compressor efficiency*
η_{turb} : *turbine efficiency*
η_{cc} : *combustion efficiency*
n_{th} : *overall thermal efficiency*
$C_{p,air}$: *air heat capacity (kJ/kgK)*
$C_{p,fuel}$: *fuel heat capacity (kJ/kgK)*
k_{air}: *air heat capacity ratio*
k_{flue}: *flue gas heat capacity ratio*
T_i: *component temperatures* (°C)
$\dot{m}_{air}$: *air mass flowrate (kg/s)*
$\dot{m}_{fuel}$: *fuel mass flowrate (kg/s)*
ΔP_{comb} : *combustion pressure loss (kPa)*

LHV: *fuel lower heat value (kJ/kg)*
$\dot{W}_{comp}$: *compressor power input (kW)*
$\dot{W}_{turb}$: *turbine power output (kW)*
$\dot{W}_{net}$: *net power output (kW)*
$\dot{Q}_{in}$: *combustion heat input (kW)*
SFC: *specific fuel consumption (kg/kWh)*
HR: *heat rate (kJ/kWh)*
HR': *time-averaged heat rate (kJ/kWh)*
n'_{th}: *time-averaged overall thermal efficiency*

Chapter 4
Multiobjective Optimization for Waste Heat Recovery of an Industrial Gas Turbine using Extreme Value Stochastic Engines

ABSTRACT

Stochastic engines or random number generators are commonly used to initialize metaheuristic approaches. This chapter discusses the incorporation of extreme value distributions into stochastic engines to improve the performance of optimization techniques when solving the complex four-objective GT-AC optimization problem.

OVERVIEW

The differential evolution (DE) approach is employed to tackle this problem. Using two extreme value distributions, two DE variants are developed by modifying their stochastic engines: Pareto-DE and Extreme-DE. The algorithms are then applied to optimize a complex MO Gas Turbine – Absorption Chiller system.

As engineering systems become more complex, there is a growing need for advanced and innovative techniques for optimizing these systems.

DOI: 10.4018/978-1-7998-1710-9.ch004

Metaheuristics have been widely used to tackle such engineering problems – especially in industries related to power generation (Marmolejo *et al.*, 2017). MO settings are also becoming commonplace in engineering; where the engineer has to consider multiple target objectives when making critical decisions. MO optimization problems could be broadly divided into two classes. The first is bi-objective problems which have lower complexity as compared to its many-objective counterparts. These problems are endowed with only two objectives and there are well established methods for solving them: Non-Dominated Sorting Genetic Algorithm (NSGA-II) (Sadeghi *et al.*, 2014) and Strength Pareto Evolutionary Algorithm (SPEA-2) (Maheta and Dabhi, 2014). As for problems with many objectives, scalarization approaches are among the most effective strategies. Examples of scalarization approaches are the weighted sum (Yang *et al.*, 2013) and the Normal-Boundary Intersection (NBI) (Charwand *et al.*, 2015; Brito *et al.*, 2014) methods. Using scalarization approaches, the multiple objectives are aggregated and the problem is transformed to a single-objective problem. This reduction in complexity then makes the problem easier to solve.

In engineering and other real-world applications, the penalty factor approach is also employed for dealing with situations with multiple objectives. Similar to the weighted sum approach, this method involves the weighted aggregation of the objectives. The distinction with the weighted sum approach is that the penalty factor method converts all the objectives into financial terms forming a single cost function. For instance, in Sheng *et al.*, (2013), an optimization of a distributed generation (DG) power system was carried out using evolutionary algorithms. The penalty factor method was used as a basis to solve the problem; where the DG utilization was maximized while minimizing the system's losses and environmental pollution. Another application using this method could be seen in the works of Daryani and Zare, (2016). In that work, the authors used the Modified Group Search Algorithm (MGSA) in tandem with the penalty factor approach to solve a MO problem in power and emission dispatch.

The Pareto distribution is among the first non-Gaussian distributions encountered in statistics. In economics this distribution shows large amounts of wealth are owned by a smaller percentage of individuals in any society. The idea is often expressed as the 80-20 Rule; 20 percent of the population owns 80 percent of the wealth (Sanders, 1987). Ever since its appearance, the Pareto distribution has found diverse applications – stretching out to other applications besides economics. In quantum statistics, the variant of Pareto distributions has been employed to study particle distributions (Biró *et al.*, 2015). These distributions have also seen application in stochastic physical processes; particularly in sub-recoil laser cooling (Bardou, 1995). It has also been used to design and assess software reliability models using failure data (Faqih, 2013; Karagrigoriou and Vonta, 2014). In Fernández *et al.*, (2016), Pareto distributions were used to improve the signal quality of radar systems. In that work the Pareto distribution coupled with a neural network was employed to estimate and eliminate background echoes in radar signals known as 'sea clutter'. Similarly in Lenz (2016), a variant of the Pareto distribution called the Generalized Pareto Distribution was employed in the field of microscopy. In that application, the distribution became critical for the construction of an efficient autofocus system.

As with the GPD, the Generalized Extreme Value (GEV) distribution has been utilized in a wide range of applications (Marmolejo and Rodriguez, 2015). For instance in electromagnetic systems, Orjubin (2007) successfully modeled the field in a reverberation chamber. The author used the GEV distribution to determine the maximum field in the chamber. GEC has also been applied to communications technology. In El-Sallabi *et al.*, (2014), the authors employed the GEV to model the radio channel parameter for improving radio channel characterization - showing the significance of extreme values in their application. In power systems, the wind energy potential was assessed using GEV (Ayuketang Arreyndip and Joseph, 2016). In that work, the GEV distribution was employed to analyze the seasonal variation of wind energy potential in Debuncha region, South-West Cameroon. As previously mentioned, extreme value distributions are specifically very handy when dealing with freak probabilistic situations resulting from complex systems. Due to the complex nature of weather systems, GEV has found much application in climate modeling. In the work of Soukissian and Tsalis, (2015), GEV was employed to predict extreme wind speeds. Accurate wind speed prediction is critical when it comes to preparing or managing natural disasters. In Cannon *et al.*, (2009), the GEV distribution was coupled with an artificial neural network to model precipitation using non-stationary and complex climatic data. GEV has also been seen to have applications in risk and reliability engineering. In Hirose (1996), the frequency of dielectric voltage breakdown was studied. The authors of that work used the GEV distribution to effectively approximate the voltage breakdown probability.

In this work a differential evolution (DE) metaheuristic in tandem with a weighted sum approach is used to optimize a combined Gas Turbine -

Absorption Chiller (GT-AC) system. For a more comprehensive details, the reader is referred to the authors past work (Ganesan et al., 2018). This GT-AC system has four target objectives; GT thermal efficiency, GT heat rate, the absorption chiller's Coefficient of Performance (COP) and the refrigeration load refrigeration load (Q_E). Due to the high complexity of the problem, the DE technique is modified to include extreme value stochastic engines. The Generalized Pareto Distribution and the Generalized Extreme Value Distribution are used to modify the DE approach yielding two techniques; the Generalized Pareto Distribution DE (Pareto-DE) and Generalized Extreme Value DE (Extreme - DE). The results of the application of the two techniques to the GT-AC system are obtained. Additionally the performances of the two new techniques are compared to the conventional DE approach.

The problem formulation given here is similar to the one discussed in Chapter 3 except for the fourth objective function; refrigeration capacity:

$$Max \rightarrow \text{Overall Thermal Efficiency, } n'_{th}$$

$$Min \rightarrow \text{Heat Rate, } HR'$$

$$Max \rightarrow \text{Coefficient of Performance, } COP'$$

$$Max \rightarrow \text{Refrigeration Capacity, } Q'_e \tag{4.1}$$

EXTREME VALUE STOCHASTIC ENGINES

Stochastic engines (or the random number generator segment) of metaheuristic algorithms have been seen to play an important role in the computational process – significantly influencing the quality of the achieved optimal solution. These stochastic engines usually generate random numbers produced from Gaussian or Normal distribution functions. In past works, it has been shown that stochastic engines which deviate from the Gaussian could improve the overall optimization results; especially for highly complex problems (Ganesan *et al.*, 2016). Such deviations from the median of standard probability distributions are known as Extreme Value Theory (Bensalah, 2000; Embrechts *et al.*, 1999). This theory explores the outcomes of events from an extreme probabilistic approach.

In most real-world optimization problems, the objective landscape is largely unknown. Similarly, the existence of the global optima could not be determined. Conventional metaheuristics assume that a Gaussian random number generator is sufficient to begin the search process. However, for extremely complex objective landscapes this assumption may breakdown. A Gaussian random number generator may not be able to place the start point of the algorithm at best place in the beginning of the search. This bad initialization could then result in a flawed or inefficient search for optimum values by the algorithm.

The Generalized Pareto distribution is a generic version of the conventional Pareto distribution originally conceived in economics as an extreme value (or non-Gaussian) distribution. In this work, the Generalized Pareto Distribution was utilized as a stochastic engine for the DE approach (Pikands, 1975; Del Castillo 2009). The cumulative distribution function for the Generalized Pareto Distribution, *P* for the random variable, $x \in X$ is given as follows:

$$P(x \in X; k, \sigma) = \begin{cases} 1 - \left[1 - (kx / \sigma)\right]^{1/k}; & k \neq 0 \quad \sigma > 0 \\ 1 - e^{-x/\sigma} \quad ; & k = 0 \quad \sigma > 0 \end{cases} \tag{4.2}$$

where σ and *k* are the scale and shape factors. In this work we consider $k = 0$; thus we take the lower equation of the piece-wise representation of equation (4.2).

The Generalized Extreme Value Distribution is a combination of three extreme value functions; Gumble, Weibull and Fréchet distributions (Walshaw, 2013). The cumulative distribution function for the Generalized Extreme Value Distribution function used in this work, *G* for the random variable, $x \in X$ is given as follows:

$$G(x \in X; \mu, \sigma) = \exp\left[-\exp\left(\frac{-x - \mu}{\sigma}\right)\right] \tag{4.3}$$

where σ and μ are the scale and location parameters respectively. In this work we consider $\sigma = \mu = 1$.

EXTREME VALUE DIFFERENTIAL EVOLUTION

In this work the Generalized Pareto and Generalized Extreme Value Distributions are used as stochastic engines in the DE technique. This modification results in the formation of two algorithms; the Generalized Pareto Distribution DE (Pareto-DE) and Generalized Extreme Value DE (Extreme - DE). Figure 4.1 depicts the components constructing this modified DE framework.

As seen in Figure 1, the two extreme value distributions (GPD and GEV) are supplemented into the existing DE algorithm. The standard random number generator (Gaussian) is replaced using the extreme value distribution. Other components of the DE algorithm (fitness function, selection and mutation) are maintained. This modification varies the DE technique's search capability aiming to improve its propensity in reaching the optima. The same procedure could be replicated on any other metaheuristic approach.

COMPUTATIONAL OUTCOME AND FINAL REMARKS

The engineering problem tackled in this work involves four nonlinear objective functions along with the required design and operational constraints. The computational strategies were applied to the problem using the weighted-sum approach as the basis. The weighted sum approach simplifies the problem reducing it to a single aggregate objective function.The aggregate objective function obtained using the mentioned approach utilizes the following inner product:

$$F = \begin{pmatrix} w_1 & w_2 & w_3 & w_4 \end{pmatrix} \begin{pmatrix} n'_{th} \\ HR' \\ COP' \\ Q'_e \end{pmatrix} : \forall w_i \in (0,1) \tag{4.4}$$

The second and fourth objectives, *HR* and Refrigeration Capacity, Q'_e have values in the scale of 10^3. To avoid a biased evaluation besides misrepresentation of the aggregated objectives, both objective functions are normalized. The normalization used the high *HR* value of 8000 kJ/kWhr

Figure 1. Extreme value differential evolution

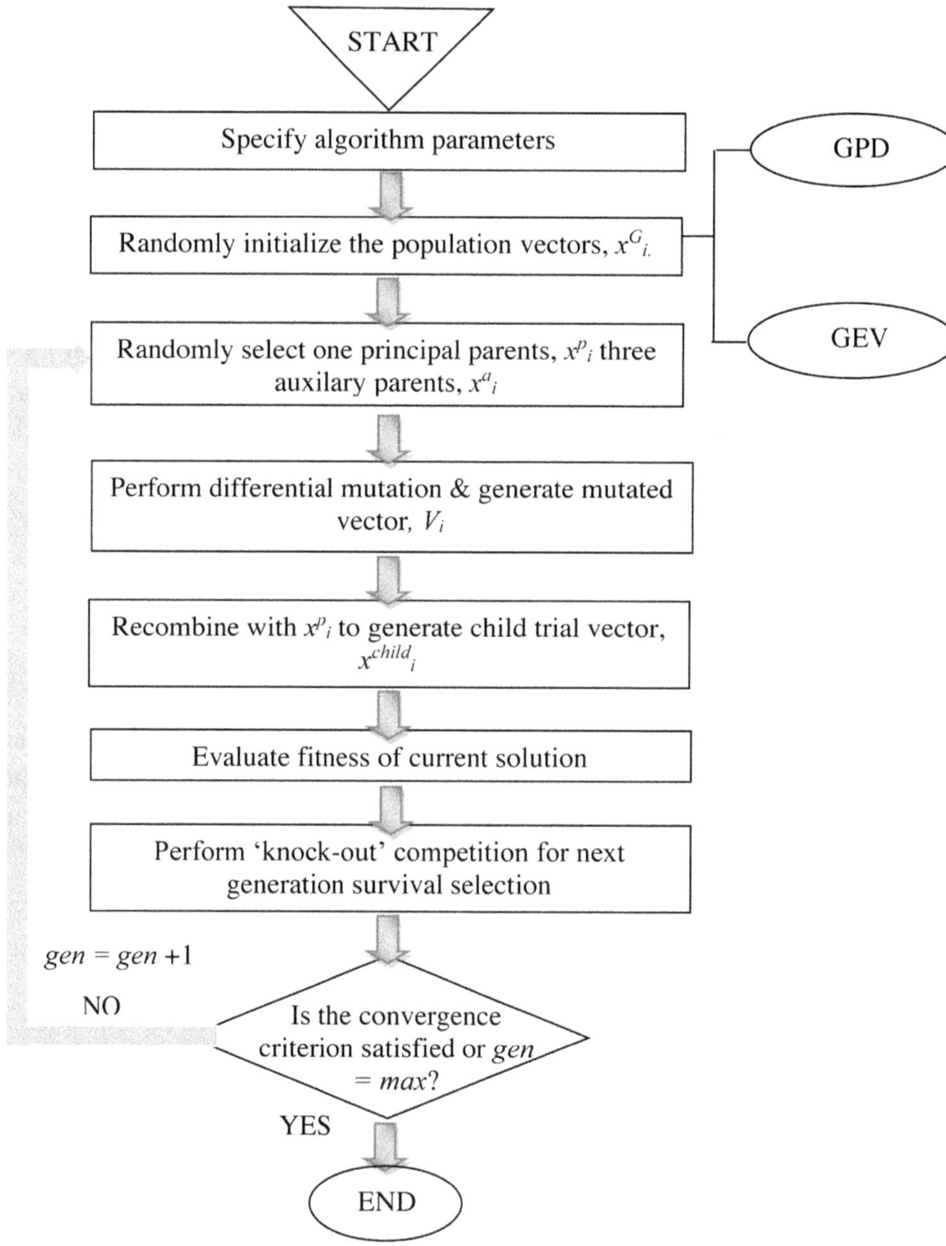

and Refrigeration Capacity, Q_e of 2000. The following equations depict this normalization procedure:

$$HR' = HR' / 8000 \tag{4.5}$$

$$Q_e' = Q_e' / 2000 \tag{4.6}$$

The optimization is carried out such that the best chiller design parameters are selected - which caters accordingly to the GT air inlet cooling requirement. The refrigeration capacity Q_e' of the chiller is maximized such that the design constraints and other requirements are not violated. This way the chiller would be engineered in a cost-efficient manner. In this work, the DE technique is modified using two extreme value distributions (Generalized Pareto Distribution and Generalized Extreme Value Distribution) yielding two algorithms the Pareto-DE and Extreme–DE respectively. As a benchmark, the conventional DE using a standard Gaussian stochastic engine (termed Gauss-DE) is applied to the problem. This way the three approaches, Pareto-DE, Extreme-DE and Gauss-DE are analyzed and compared. The concept of solution dominance is employed to evaluate the solution quality (Ganesan *et al.*, 2013; Ganesan *et al.*, 2015). For this evaluation, the hypervolume indicator (HVI) is employed with the nadir point $(0.1, 1, 0.1, 0.1)$ (Emmerich *et al.*, 2005).

Individual solutions obtained at various scalar values were used to construct the Pareto frontier. Take note that the Pareto–DE technique and Pareto frontier sound similar but in reality are two distinct concepts. The Pareto-DE is a technique endowed with a stochastic engine which randomly initializes the algorithm based on the Generalized Pareto Distribution. On the other hand the Pareto frontier is way to depict and analyze solutions of MO problems. Executing randomized techniques such as evolutionary algorithms require multiple program executions. In this work, each individual solution was obtained after 5 independent runs and the best solution is selected for analyses. The GT-AC system is modeled as dynamic system running at a 24-hour cycle. Due to this the objective values and other parameters are hourly time-averaged. The individual solutions obtained are ranked for each Pareto frontier (best, median and worst).

All techniques employed in this work were designed using the C++ programming language. The software was developed and ran on a personal computer (PC) with an Intel® Core ™ i7 – 3520M processor operating at 2.9 GHz. The ranked solutions obtained using the Gauss-DE approach are given in Table 1. The related Pareto frontier is depicted in three scatter plots in Figure 2:

The best, median and worst solutions achieved by the Gauss-DE are taken at the weights (0.3, 0.3, 0.3, 0.1), (0.3, 0.4, 0.2, 0.1) and (0.5, 0.1, 0.1, 0.3) respectively. Table 2 and Figure 3 gives the ranked solutions and the Pareto frontier obtained using the Extreme – DE.

The associated weights for the ranked individual solutions obtained using the Extreme-DE are best: (0.1 0.4, 0.3, 0.2), median: (0.1, 0.7, 0.1, 0.1) and worst: (0.5, 0.3, 0.1, 0.1). The ranked solutions obtained using the Pareto-DE approach is given in Table 3 while Figure 4 depicts the associated Pareto frontier. The weight values for the ranked individual solutions; best, median and worst obtained using the Pareto-DE are (0.3, 0.2, 0.4, 0.1), (0.1, 0.1, 0.2, 0.6) and (0.3, 0.1, 0.5, 0.1) respectively.

The most dominant individual solution was achieved by the Pareto-DE followed by the Gauss-DE and the Extreme-DE respectively. The dominance levels were measured using the HVI. The individual solutions produced by the Pareto-DE outrank that of the Extreme-DE and Gauss-DE by 130.787% and 102.63% respectively. It is observed that in terms of individual solution,

Table 1. The ranked individual solutions obtained using the Gauss-DE technique

Description		Best	Median	Worst
Objective Functions	n'_{th}	0.34322	0.337385	0.336577
	HR'	0.711423	0.723731	0.725469
	COP'	1.94784	1.94352	1.93796
	Q'_e	0.754	0.53105	0.500247
Decision Variable	t_a	37.2475	36.4172	35.6108
	t_e	7.64611	7.5598	7.61699
	t_g	486.461	485.598	486.17
	t_c	37.3539	37.4402	37.383
	E_L	0.564611	0.55598	0.561699
	M_a	105.709	104.015	103.781
	M_g	105.709	104.015	103.781
	W_{net}	137.288	134.954	134.631
	ΔT	11.3901	8.01133	7.54347
Metric	HVI	84.821	52.114	47.778

Figure 2. Pareto frontier constructed using Gauss-DE

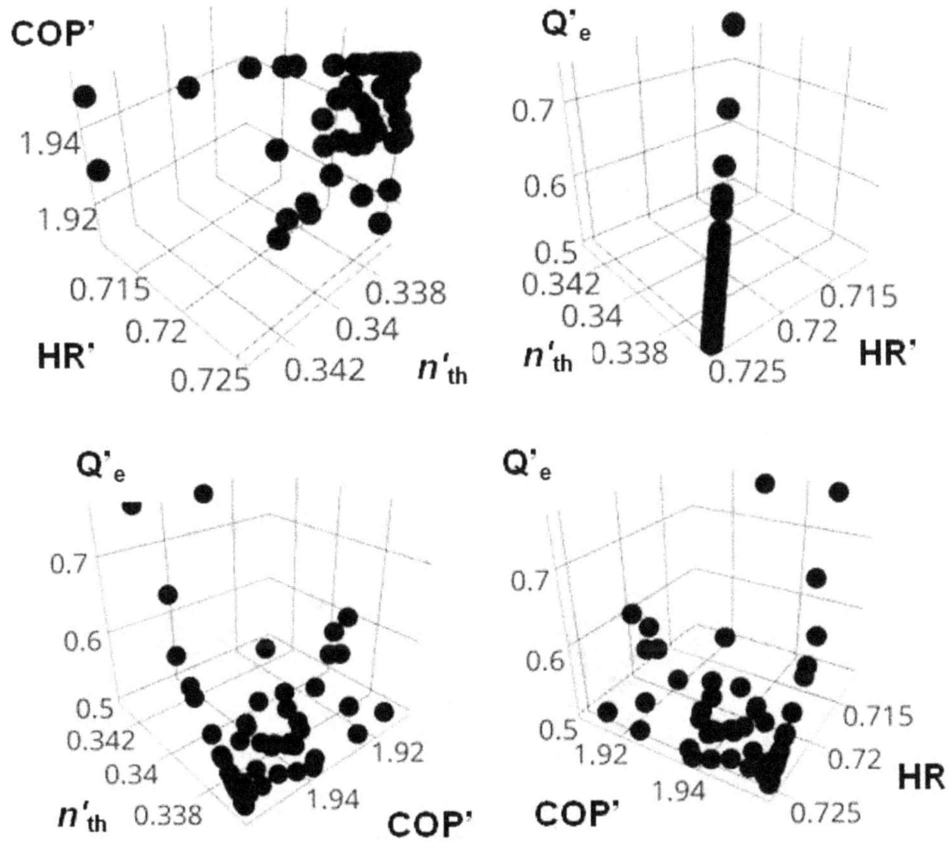

the Gauss-DE outweighs the Extreme-DE. The overall frontier dominance attained by all three approaches is given in Figure 5:

Similarly it can be seen in Figure 5 that the overall frontier dominance is led by the Pareto-DE technique. Although the frontier produced by the Extreme-DE outweighs the Gauss-DE, the difference is very small (approximately 4.516%). The Pareto-DE on the other hand leaped far ahead from the other two approaches in terms of frontier dominance with the following differences:

$$\text{Extreme-DE} \approx 58.531\% \tag{4.7}$$

$$\text{Gauss-DE} \approx 65.691\% \tag{4.8}$$

Figure 3. Pareto frontier constructed using Extreme-DE

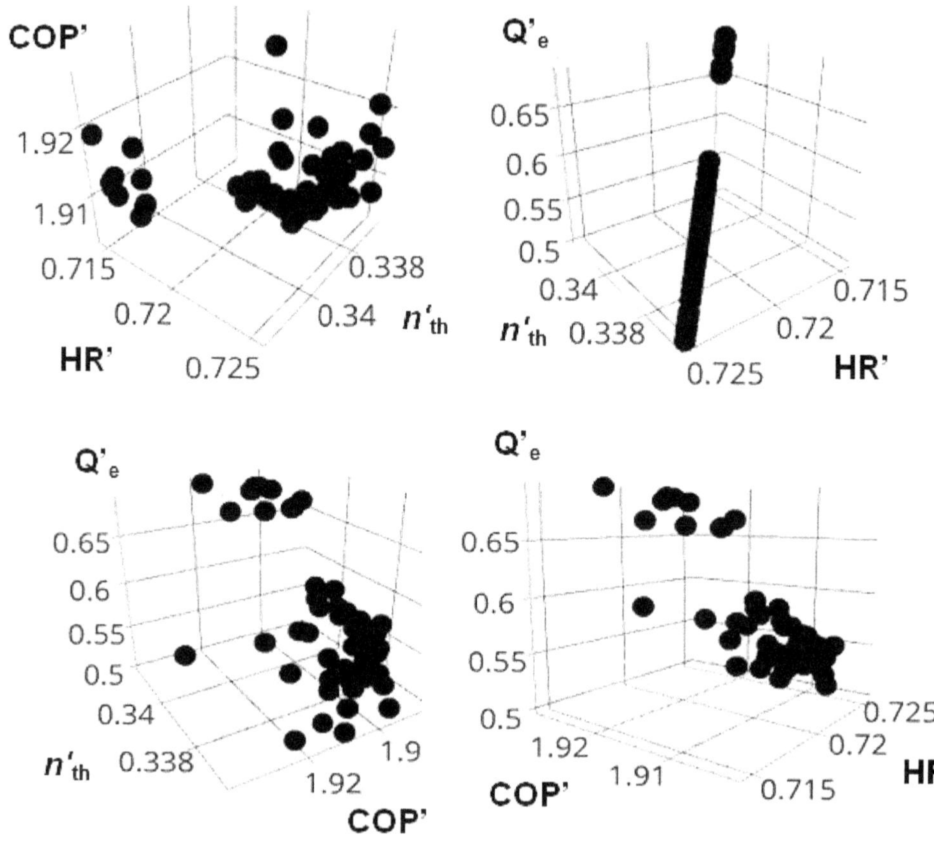

The computational time taken for the Pareto –DE to construct the entire Pareto frontier was 3.765 seconds. The Extreme-DE and Gauss-DE produced the frontiers in 61.147 and 4.875 seconds respectively. Note that the Extreme-DE took the longest computational time; more than a minute.

In Figure 2, it can be seen that the solution points on the frontier generated using the Gauss-DE are concentrated on certain regions in the objective space. The solutions are particularly dense at areas with high COP, low n_{th}, high *HR* and low Q_e (indicated by the circles on Figure 2).The same goes for the Pareto frontier produced by the Extreme-DE (Figure 3). In this case the solutions are conglomerated in two different sections of the objective space – similarly marked by circling the regions on Figure 3. The first section, have solutions with dispersed COP values while having low *HR*, high Q_e and high n_{th} values. As for the second section, the solution points have well dispersed n_{th} and *HR*. Nevertheless these points center on regions with high

Table 2. The ranked individual solutions obtained using the Extreme-DE technique

Description		Best	Median	Worst
Objective Functions	n'_{th}	0.341698	0.338127	0.336598
	HR'	0.714591	0.722142	0.725423
	COP'	1.91877	1.90537	1.91441
	Q'_e	0.69358	0.5589	0.50072
Decision Variable	t_a	33.1331	31.1403	32.3352
	t_e	8.01339	8.10903	7.94516
	t_g	490.134	491.09	489.452
	t_c	36.9866	36.891	37.0548
	E_L	0.601339	0.610903	0.594516
	M_a	105.267	104.231	103.787
	M_g	105.267	104.231	103.787
	W_{net}	136.679	135.251	134.639
		10.5092	8.44117	7.55578
Metric	*HVI*	74.47281	54.81707	47.23357

Table 3. The ranked individual solutions obtained using the Pareto-DE technique

Description		Best	Median	Worst
Objective Functions	n'_{th}	0.356412	0.342766	0.337216
	HR'	0.685084	0.712365	0.724094
	COP'	1.94309	1.94373	1.94837
	Q'_e	1.25486	0.736584	0.524627
Decision Variable	t_a	37.0995	36.6551	37.0882
	t_e	8.04374	7.69218	7.48779
	t_g	490.437	486.922	484.878
	t_c	36.9563	37.3078	37.5122
	E_L	0.604374	0.569218	0.548779
	M_a	109.537	105.577	103.966
	M_g	109.537	105.577	103.966
	W_{net}	142.565	137.106	134.886
		19.0287	11.1273	7.91349
Metric	*HVI*	171.8732	81.95635	51.36895

Figure 4. Pareto frontier constructed using Pareto-DE

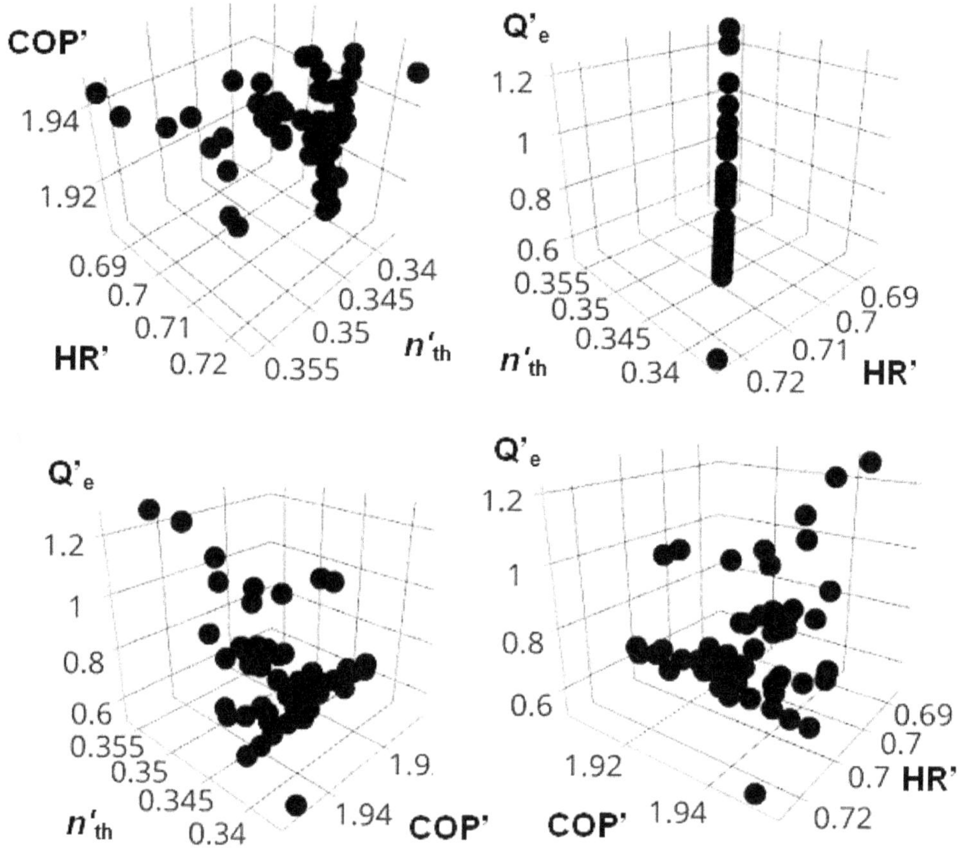

Figure 5. Dominance ranking of the generated Pareto frontiers

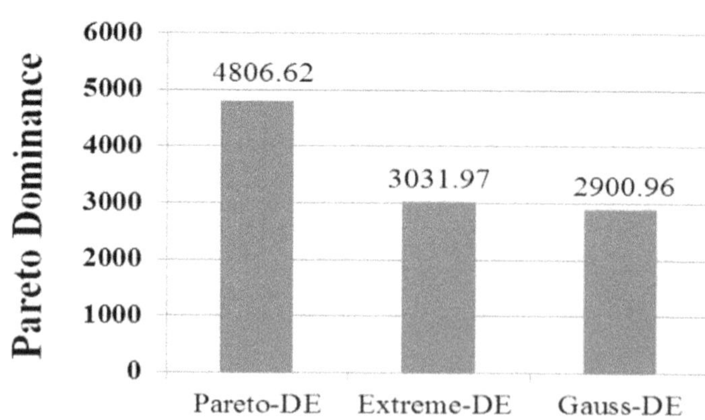

COP and low Q_e. The most well spread Pareto frontier was produced by the Pareto-DE technique (portrayed in Figure 4). As the technique approximates the true frontier, the level of frontier dominance would increase. A diversely spread frontier would cover most of objective space closely estimating the true Pareto frontier. This is why the Pareto-DE has a diversely spread and most dominant frontier as compared to the other approaches employed in this work. The Gauss-DE produced the worst frontier since its solutions are narrowly focused on a single region. Although not as diverse as the Pareto-DE, the Extreme-DE could sufficiently diversify its Pareto frontier - not limiting its solution spread to a single region in the objective space. This diversification enables it to outweigh the overall frontier dominance of the solutions produced by the Gauss-DE.

Therefore extreme value stochastic engines do influence the solution quality. However, the selection of the right extreme value distribution is critical for empowering the search capability of the metaheuristic. Execution time is a seminal criterion when evaluating algorithmic performance. The computational effort undergone by the three DE variants reflect the Pareto dominance achieved by their respective frontiers. The Pareto-DE approach sweeps through the objective space smoothly, diversely spreading its solution throughout. Thus, the Pareto-DE technique does experience stagnation at any region in the objective space. This may contribute to the speedy convergence taking up minimal execution time as compared to the Extreme-DE and Gauss–DE approaches. The Gauss-DE comes second after the Pareto-DE approach. The Gauss-DE stagnates at only one region of the objective space. Due to its limited spread it completes its exploration swiftly bringing the program to a halt. Albeit the lengthy execution time of the Extreme-DE, its search covers a wider area as compared to the Gauss-DE. The sacrifice in computational efficiency is too costly in comparison with the Pareto-DE algorithm.

The only differences in the techniques employed in this work are their respective stochastic engines. Thus the algorithmic complexity of the techniques utilized in this work does not differ much from one another. All the DE variants performed stable computations during program executions. The solutions generated did not violate any of the defined constraints and thus remain feasible.

Based on the numerical experiments performed, it was observed that the stochastic engines do significantly influence the optimization results. The results in this case were gauged using the level of dominance measured using the HVI. Nevertheless the type of stochastic engine used also heavily affects the optimization results. As seen in the previous section, the frontier produced by the Pareto-DE outweighs that of the Extreme-DE and Gauss-DE by a large margin. On the other hand, when it comes to individual solutions, the Gauss-DE outranks the Extreme-DE. This shows that the wrong choice of stochastic engine could reduce algorithmic performance to such an extent that it could produce results which are way worse than conventional approaches (e.g. Gauss-DE). The Pareto-DE effectively solved the complex real-world MO optimization problem: the combined GT-AC system.

It was found that different stochastic engines differently influence the optimization results when applied to more complex problems (four-objective MO problems). In this chapter, the factors of the probability distribution (shape, scale and location) were fixed. The effect of these factors on the algorithmic performance could also be analyzed. This would be particularly useful for tackling real-world problems which are highly complex. Additionally, future works could also investigate the mathematical aspects of using specific extreme-value distribution to tackle certain classes of optimization problems. This way the mathematical enhancement in the metaheuristic would exactly match the attempted problem. Further studies into the mathematical foundations of extreme value distributions would help provide useful insights in designing extreme value metaheuristics (Nakajima *et al.*, 2017; Ning and Bloomfield, 2017). Other optimization algorithms could also be embedded to find for optimal factors for the distributions in the stochastic engines. The strategy of switching stochastic engines could also be implemented using other evolutionary approaches besides DE. This way the extreme value distributions may also be tested on metaheuristics such as particle swarm optimization (PSO) (Singh *et al.*, 2015), Bacteria Foraging Optimization (Nasiruddin *et al.*, 2015), Gravitational Search Algorithm (Ganesan and Elamvazuthi, 2017) and other bio-inspired strategies (Plucar *et al.*, 2017).

REFERENCES

Ayuketang Arreyndip, N., & Joseph, E. (2016). Generalized Extreme Value Distribution Models for the Assessment of Seasonal Wind Energy Potential of Debuncha, Cameroon. *Journal of Renewable Energy*.

Bardou, F. (2005). Fractal time random walk and subrecoil laser cooling considered as renewal processes with infinite mean waiting times. In *Chaotic Dynamics and Transport in Classical and Quantum Systems* (pp. 281–301). Springer Netherlands. doi:10.1007/1-4020-2947-0_12

Bensalah, Y. (2000). *Steps in applying extreme value theory to finance: a review*. Vancouver: Bank of Canada.

Charwand, M., Ahmadi, A., Heidari, A. R., & Nezhad, A. E. (2015). Benders decomposition and normal boundary intersection method for multiobjective decision making framework for an electricity retailer in energy markets. *IEEE Systems Journal*, *9*(4), 1475–1484. doi:10.1109/JSYST.2014.2331322

Daryani, N., & Zare, K. (2016). *Multiobjective power and emission dispatch using modified group search optimization method. Ain Shams Engineering Journal*.

Del Castillo, J., & Daoudi, J. (2009). Estimation of the generalized Pareto distribution. *Statistics & Probability Letters*, *79*(5), 684–688. doi:10.1016/j.spl.2008.10.021

El-Sallabi, H., Abdallah, M., Chamberland, J. F., & Qaraqe, K. (2014, December). A statistical model for delay domain radio channel parameter affected with extreme values. In *Antennas and Propagation (ISAP), 2014 International Symposium on* (pp. 165-166). IEEE. 10.1109/ISANP.2014.7026582

Embrechts, P., Resnick, S. I., & Samorodnitsky, G. (1999). Extreme value theory as a risk management tool. *North American Actuarial Journal: NAAJ*, *3*(2), 30–41. doi:10.1080/10920277.1999.10595797

Emmerich, M., Beume, N., & Naujoks, B. (2005). An EMO Algorithm Using the Hypervolume Measure as Selection Criterion. *Conference on Evolutionary Multi-Criterion Optimization (EMO 2005)*, 62–76. 10.1007/978-3-540-31880-4_5

Faqih, K. M. (2013). The Performance of Software Reliability Models: A View Point. *International Journal of Performability Engineering*, *9*(4), 375–386.

Ganesan, T., Aris, M. S., & Vasant, P. (2018). Extreme value metaheuristics for optimizing a many-objective gas turbine system. *International Journal of Energy Optimization and Engineering*, *7*(2), 76–96. doi:10.4018/IJEOE.2018040104

Ganesan, T., Elamvazuthi, I., Shaari, K. Z. K., & Vasant, P. (2013). Multiobjective optimization of green sand mould system using chaotic differential evolution. In *Transactions on Computational Science XXI* (pp. 145–163). Springer Berlin Heidelberg. doi:10.1007/978-3-642-45318-2_6

Ganesan, T., Elamvazuthi, I., & Vasant, P. (2015). Multiobjective design optimization of a nano-CMOS voltage-controlled oscillator using game theoretic-differential evolution. *Applied Soft Computing*, *32*, 293–299. doi:10.1016/j.asoc.2015.03.016

Ganesan, T., Vasant, P., & Elamvazuthi, I. (2016). Multiobjective optimization using particle swarm optimization with non-Gaussian random generators. *Intelligent Decision Technologies*, *10*(2), 93–103. doi:10.3233/IDT-150241

Hirose, H. (1996). Maximum likelihood estimation in the 3-parameter Weibull distribution. A look through the generalized extreme-value distribution. *IEEE Transactions on Dielectrics and Electrical Insulation*, *3*(1), 43–55. doi:10.1109/94.485513

Karagrigoriou, A., & Vonta, I. (2014, September). Statistical Inference for Heavy-Tailed Distributions in Technical Systems. In *Availability, Reliability and Security (ARES), 2014 Ninth International Conference on* (pp. 412-419). IEEE. 10.1109/ARES.2014.62

Kohlenbach, P., & Ziegler, F. (2008). A dynamic simulation model for transient absorption chiller performance. Part I: The model. *International Journal of Refrigeration*, *31*(2), 217–225. doi:10.1016/j.ijrefrig.2007.06.009

Lenz, R. (2016). Generalized Pareto Distributions—Application to Autofocus in Automated Microscopy. *IEEE Journal of Selected Topics in Signal Processing*, *10*(1), 92–98. doi:10.1109/JSTSP.2015.2482949

Maheta, H. H., & Dabhi, V. K. (2014, February). An improved SPEA2 Multi objective algorithm with non dominated elitism and Generational Crossover. In *Issues and Challenges in Intelligent Computing Techniques (ICICT), 2014 International Conference on* (pp. 75-82). IEEE. 10.1109/ICICICT.2014.6781256

Marmolejo, J. A., & Rodriguez, R. (2015). Fat tail model for simulating test systems in multiperiod unit commitment. *Mathematical Problems in Engineering*.

Marmolejo, J. A., Velasco, J., & Selley, H. J. (2017). An adaptive random search for short term generation scheduling with network constraints. *PLoS One*, *12*(2). doi:10.1371/journal.pone.0172459 PMID:28234954

Orjubin, G. (2007). Maximum field inside a reverberation chamber modeled by the generalized extreme value distribution. *IEEE Transactions on Electromagnetic Compatibility*, *49*(1), 104–113. doi:10.1109/TEMC.2006.888172

Sadeghi, J., Sadeghi, S., & Niaki, S. T. A. (2014). A hybrid vendor managed inventory and redundancy allocation optimization problem in supply chain management: An NSGA-II with tuned parameters. *Computers & Operations Research*, *41*, 53–64. doi:10.1016/j.cor.2013.07.024

Sanders, R. (1987). The Pareto principle: Its use and abuse. *Journal of Services Marketing*, *1*(2), 37–40. doi:10.1108/eb024706

Sheng, W., Liu, K. Y., Liu, Y., Meng, X., & Song, X. (2013). A new DG multiobjective optimization method based on an improved evolutionary algorithm. *Journal of Applied Mathematics*.

Walshaw, D. (2013). Generalized extreme value distribution. Encyclopedia of Environmetrics.

APPENDIX

Nomenclature and Abbreviations for the GT-AC System

Subscripts

1: *stream from absorber to pump*
2: *stream from pump to solution HEX*
3: *stream from HEX to generator*
4: *stream from generator to HEX*
5: *stream from HEX to valve*
6: *stream from valve to absorber*
7: *stream from generator to condenser*
8: *stream from condenser to expansion valve*
9: *stream from expansion valve to evaporator*
1: *stream from evaporator to absorber*

Absorption Chiller

X_s: *strong solution concentration* (kg LiBr/ kg solution)
X_w: *weak solution concentration* (kg LiBr/ kg solution)
t_a: *absorber temperature* (°C)
t_e: *evaporator temperature* (°C)
t_g: *generator temperature* (°C)
t_c: *condenser temperature* (°C)
P_e: *evaporator pressure* (mm Hg)
P_c: *condenser pressure* (mm Hg)
H_i : *enthalpy of pure water* (kcal/kg)
Q_E: *transient refrigeration load* (kcal/hr)
$\dot{m}_R$: *refrigerant mass flowrate* (kg/hr)
$\dot{m}_W$: *weak LiBr solution mass flowrate* (kg/hr)
$\dot{m}_S$: *strong LiBr solution mass flowrate* (kg/hr)
E_L : *HEX effectivity*
C_{X_1} : *specific heat of the strong solution* (kcal/kg °C)
C_{X_4} : *specific heat of the weak solution* (kcal/kg °C)
Q_e : *heat input to the evaporator* (kcal)
Q_g : *heat input to the generator* (kcal)

Q_c : *heat output from the evaporator* (kcal)
Q_a : *heat output from the absorber* (kcal)
dQ_e / dt : *heat rate - input to the evaporator* (kcal/hr)
dQ_g / dt : *heat rate - input to the generator* (kcal/hr)
dQ_c / dt : *heat rate - output to the evaporator* (kcal/hr)
dQ_e / dt : *heat rate - input to the absorber* (kcal/hr)
COP: *coefficient of performance*
COP_i: *ideal/max coefficient of performance*
RPR: *relative performance ratio*
$\dot{m}_{a,t}(i)$: *outlet mass flowrate at the absorber tube* (kg/hr)
$\dot{m}_{g,t}(i)$: *outlet mass flowrate at the generator tube*(kg/hr)
$M_a(i)$: *mass stored at the absorber sump (shell side) tube* (kg)
$M_g(i)$: *mass stored at the generator sump (shell side) tube*(kg)
Δt : *simulation time-interval* (hr)
COP': *time-averaged coefficient of performance*
ΔT : *temperature differential for cooling* (hr)

Gas Turbine

p_i: *component pressures (kPa)*
r: compression pressure ratio
η_{comp} : *compressor efficiency*
η_{turb} : *turbine efficiency*
η_{cc} : *combustion efficiency*
n_{th} : *overall thermal efficiency*
$C_{p,air}$: *air heat capacity (kJ/kgK)*
$C_{p,fuel}$: *fuel heat capacity (kJ/kgK)*
k_{air}: *air heat capacity ratio*
k_{flue}: *flue gas heat capacity ratio*
T_i: *component temperatures* (°C)
$\dot{m}_{air}$: *air mass flowrate (kg/s)*
$\dot{m}_{fuel}$: *fuel mass flowrate (kg/s)*
ΔP_{comb} : *combustion pressure loss (kPa)*
LHV: fuel lower heat value (kJ/kg)
$\dot{W}_{comp}$: *compressor power input (kW)*

$\dot{W}_{turb}$: *turbine power output (kW)*
$\dot{W}_{net}$: *net power output (kW)*
$\dot{Q}_{in}$: *combustion heat input (kW)*
SFC: specific fuel consumption (kg/kWh)
HR: heat rate (kJ/kWh)
HR': time-averaged heat rate (kJ/kWh)
n'_{th}: *time-averaged overall thermal efficiency*

Chapter 5
Biofuel Supply Chain Optimization Using Lévy-Enhanced Swarm Intelligence

ABSTRACT

Supply chain planning/optimization presents various challenges to decision makers globally due to its highly complicated nature as well as its large-scale structure. Over the years various state-of-the-art methods have been employed to model supply chains. Optimization techniques are then applied to such models to help with optimal decision making. However, with highly complex industrial systems such as these, conventional metaheuristics are still plagued by various drawbacks. Strategies such as hybridization and algorithmic modifications have been the focus of previous efforts to improve the performance of conventional metaheuristics. In light of these developments, this chapter presents two solution methods for tackling the biofuel supply chain problem.

OVERVIEW

Fuel supply chains have recently picked up interest by researchers in academia as well as decision makers in various industries (Ba *et al.*, 2016; Yue *et al.*, 2014; Ghaderi *et al.*, 2016). Modeling and optimizing these large-scale fuel supply chains have deemed to be challenging. An example of such a model development could be seen in the work of Awudu and Zhang, (2013). In that

DOI: 10.4018/978-1-7998-1710-9.ch005

work, the authors proposed a stochastic planning model for a biofuel supply chain. The supply chain consisted of biomass suppliers, biofuel refinery plants and distribution centers. Targeted at optimizing profits, the Geometric Brownian Motion and Benders decomposition method with Monte Carlo simulation was applied. Besides benchmarking their stochastic model against a deterministic model, the authors performed sensitivity analysis on various parameters using their model. Another work towards modeling fuel supply chains was presented in Poudel *et al.*, (2016). There a pre-disaster planning model was developed to strengthen the biofuel supply chain system. The pre-disaster model considered the linking and selection of facilities that reflected the post-disaster connectivity and biofuel-related costs. The model was developed using the generalized Bender's decomposition algorithm and validated using industrial data from Mississippi and Alabama, United States.

On the other hand, the optimization work in Lin *et al.*, (2014) aimed to minimize the annual biomass-ethanol production costs. The supply chain problem solved in that work was a large-scale model consisting of biomass harvesting, stacking, in-field transportation, transportation, preprocessing, packing and storage, to ethanol production and distribution. Using the mixed integer programming technique, the authors managed to reduce the cost of production (biorefinery) by 62%. A similar work was carried out in Zhang *et al.*, (2013), where the switchgrass-based bioethanol supply chain was modelled and optimized. The location of study was North Dakota, United States. Similar to Lin *et al.*, (2014), the model was optimized using mixed integer linear programming; obtaining optimal usage of marginal land for switchgrass production to harvest bioethanol in a sustainable and economical manner. In Osmani and Zhang, (2014), a large-scale sustainable dual feedstock bioethanol supply chain was optimized in a stochastic environment. The optimization was carried out while considering the biomass supply, purchase price, demand and sales price. The optimization work used a mixed integer linear programming technique in tandem with a sample average approximation algorithm as a decomposition technique. Similar works using mixed integer linear/nonlinear programming approaches for optimizing a fuel supply chain could be seen in Gao and You, (2015) and Osmani and Zhang, (2013). A work that does not employ a mixed integer approach for fuel supply chain optimization could be seen in Marufuzzaman *et al.*, (2014). In that work, the solution method employed combined Lagrangian relaxation and L-shaped techniques. Their computational experiments provided key insights on carbon regulatory mechanisms and uncertainties in biofuel supply chains. A broad review on metaheuristic techniques implemented to bioenergy supply chains is given in Castillo-Villar, K.K., (2014) and De Meyer *et al.*, (2014).

Swarm intelligence has been employed extensively in various industrial and engineering scenarios (Ling *et al.*, 2016; Gao *et al.*, 2015; Divyesh *et al.*, 2017; Ganesan *et al.*, 2016). Besides, swarm-based optimization techniques have also been actively used in various types of supply chain problems. For instance in Yeh *et al.*, (2016), optimization was done on a three-stage supply chain network while considering the deterioration effect. Minimizing the total cost, the authors employed a novel swarm algorithm called the Simplified Swarm Optimization technique. The results was then compared to the standard particle swarm optimization (PSO) and genetic algorithm techniques. In Chen *et al.*, (2017), the closed loop supply chain MO problem for the solar cell industry was tackled. Using the multiobjective PSO (MOPSO), the Pareto frontier of the large-scale supply chain problem was effectively attained. Another interesting work involving the application of swarm intelligence in supply chains could be seen in Kumar *et al.*, (2016). In that work, the vehicle routing problem (production and pollution) was solved using the self-learning PSO algorithm. Since the problem was MO in nature, the results obtained using the self-learning PSO was compared with the NSGA-II approach.

In addition to PSO-based algorithms, cuckoo search (CS) has also been effective in solving real-world supply chain problems. This can be seen in the work of Mattos *et al.*, (2017) – where a series of metaheuristics as well as CS was applied to a real-world supply chain (consumer packaged goods industry). Rigorous comparative analysis was carried out on the performance as well as the results produced by the techniques employed in that work. Similarly in Abdelsalam and Elassal, (2014), CS was employed among other metaheuristics to solve the joint economic lot-sizing problem. The problem was for a multi-layered supply chain with multi-retailers and a single manufacturer/supplier. Upon comparing the computational results, the CS technique was shown to favor the centralized policy as opposed to a decentralized one for safety stock. Another application of CS in supply chains is given in the work of Srivastav and Agrawal, (2017). In that work CS was employed to solve a MO lot-size reorder point backorder inventory problem. The triple-objective problem consisted of the following targets (annual): total relevant cost, expected number of stocked out units and expected number of stocked out occasions. The CS technique managed to generate an efficient Pareto curve between cost and service levels for the decision-makers.

BIOFUEL SUPPLY CHAIN MODEL

The fuel supply chain model framework employed in this chapter was developed in Tan *et al.*, (2017). In that work only two objective functions were considered: profit (Pr) and social welfare (SW). The environmental benefits objective was incorporated into the SW function. In this chapter, the environmental benefits were considered as isolated from the SW function - taken as an independent objective (denoted Env). Therefore the fuel supply chain model in this chapter consists of three objective functions to be maximized followed by inequality constraints (see equations (5.6 - 5.15). The objective functions are shown in equations (5.1 - 5.5):

$$\mathrm{Pr} = P(1-EC)\times\sum_t q_t - \left\{FCp + \sum_t\left[GC\cdot q_t + \sum_i\left(SC\cdot IQ_{i,t} + \sum_k SQ_{i,k,t}\cdot PP_i\right) + Y_{1t}\cdot extraY_1 + Y_{2t}\cdot extraY_2\right]\right\} \tag{5.1}$$

$$SW = ACS\cdot(1-EC)\sum_t q_t + GT - GS\cdot(1-EC)\sum_t q_t \tag{5.2}$$

$$Env = AC\cdot\left[CET\cdot(1-EC)\cdot\sum_t q_t - \left(CEcb - CEtp\right)\right] \tag{5.3}$$

such that,

$$CEcb = 2\sum_{i,k,t} PQ_{i,k,t}\cdot\sum_{i,k}\left[\frac{CEncb_{i,k}\cdot Dcb_{i,k}}{LCcb_{i,k}}\right] + \left(\sum_{i,k,t} PQ_{i,k,t}\cdot\sum_{i,k} CEicb_{i,k} Dcb_{i,k}\right) \tag{5.4}$$

$$CEtp = 2\sum_{i,k,t} SQ_{i,k,t}\cdot\sum_{i,k}\left[\frac{CEntb_{i,k}\cdot Dbp_{i,k}}{LCtp_{i,k}}\right] + \left(\sum_{i,k,t} PQ_{i,k,t}\cdot\sum_{i,k} CEicb_{i,k} Dcb_{i,k}\right) \tag{5.5}$$

The constraints for the biofuel supply chain model are given below:

$$q_t \leq q_{\max} \tag{5.6}$$

$$IQ_{i,t} \geq SIlb_i \tag{5.7}$$

$$\sum_i IQ_{i,t} \leq IQ_{\max} \tag{5.8}$$

$$Q_{\min} \leq \sum_t q_t \leq Q_{\max} \tag{5.9}$$

$$HV_{\min} \leq \sum_i HV_i \cdot BR_{i,t} \leq HV_{\max} \tag{5.10}$$

$$\sum_i SQ_{i,k,t} \geq SQ_{\min i,k} \tag{5.11}$$

$$\sum_i PQ_{i,k,t} \leq PQ_{\max} \tag{5.12}$$

$$\sum_i PQ_{i,k,t} \leq AQ_{\max,i,t} \tag{5.13}$$

$$WR_{i,k,t} \leq \left[\frac{1 - MCori_{i,t}}{1 - MC_{\max.i,t}}\right] \tag{5.14}$$

$$\sum_{i,t} SQ_{i,k,t} \cdot PP_{i,k,t} \geq \begin{bmatrix} FCb_k + \sum_{i,t} SQ_{i,k,t} \cdot \left(AP_{i,k,t} + \frac{TCcb_{i,k} \cdot Dcb_{ik}}{LCcb_{i,k}} \right) \frac{1}{WR_{i,k,t}} \\ + \sum_{i,t} SQ_{i,k,t} \cdot \frac{TCtp_{i,k} Dbp_k}{LCtp_{ik}} \end{bmatrix} \cdot \left(1 + ER_k\right) \tag{5.15}$$

such that,

$$i \in [1,2],\ k \in [1,10],\ t \in [1,12] \tag{5.16}$$

The decision variables are: q_t, $IQ_{i,t}$, $SQ_{i,k,t}$, $PP_{i,k,t}$, $PQ_{i,k,t}$ and $BR_{i,t}$. Details on the parameter settings of the model used in this chapter are as specified in Tan *et al.*, (2017).

SWARM-BASED TECHNIQUES

Cuckoo Search

Similar to swarm intelligence techniques, CS is a population based stochastic search and optimization algorithm (Mareli and Twala, 2017; Joshi *et al.*, 2017). CS was originally inspired by brood parasitism often found among some species of cuckoo birds. The parasitism occurs when the cuckoo birds lay their eggs in the nests of other bird species (non-cuckoo birds). In a CS algorithm, a search pattern corresponds to a nest - where an individual attributed to a search pattern is represented by a cuckoo egg. Thus the cuckoo egg symbolizes a potential solution candidate to the optimization problem; which will be tested by the fitness function. Then the solution will be accounted for in subsequent iterations to see if it is fit (fulfilling the fitness criteria). Otherwise the solution (egg) would be discarded and another candidate solution would be considered. This way the solution to the optimization problem would be iteratively improved as the technique explores the search space. The stochastic generator of the CS technique is based on the heavy-tailed random walk probability distribution; Lévy flights. The equation for the iterative candidate solution *i* and iteration *t* for the CS algorithm is given as:

$$y_i^{t+1} = y_i^t + \beta \cdot Levy(\lambda) \tag{5.17}$$

such that the Lévy distribution is given as follows:

$$Levy(\lambda) = t^{-\lambda} \tag{5.18}$$

where t is the random variable, $\beta > 0$ is the relaxation factor (which is modified based on the problem at hand), y_i^t is the candidate solution and $\lambda \in (1,3]$ is the Lévy flight step-size. With $t \geq 1$, λ is related to the fractal dimension and the Lévy distribution becomes a specific sort of Pareto distribution. The CS algorithm is based on a few fundamental philosophies. For instance each cuckoo bird lays a single egg at one time and randomly places the egg in a selected nest. The second being: via fitness screening, the best egg (candidate solution) is carried forward into the next iteration. The worst solutions are discarded from further iterations. The nests represent the objective space (or the optimization problem landscape). The parameter setting for the CS technique used in this chapter is shown in Table 5.1 while its respective algorithm is given in Algorithm 5.1.

Algorithm 5.1: Cuckoo Search (CS)

Step 1: Initialize algorithmic parameters; y_i, β, λ, N
Step 2: Define parameters in the constraints and decision variable
Step 3: Via Lévy flights randomly lay a cuckoo egg in a nest
Step 4: Define fitness function based on solution selection criteria
Step 5: Screen eggs and evaluate candidate solution

Table 1. CS Settings

Parameters	Values
Total Number of Eggs (N)	20
Number of nests, *nests*	4
Lévy flight step-size, λ	1.5
Relaxation factor, β Maximum number iteration, T_{max}	0.8 300

```
        IF: fitness criteria is satisfied
                Select candidate solution (egg) to be
considered in the next iteration, n+1
        ELSE: fitness criteria is not satisfied
                Discard candidate solution (egg) from further
iterations
Step 6: Rank the best solutions obtained during fitness
screening
Step 7: If the fitness criterion is satisfied and t= T_max  halt
and print solutions, else proceed to Step 3
```

GRAVITATIONAL SEARCH ALGORITHM

The GSA algorithm is a metaheuristic algorithm inspired by concepts from Newtonian gravitation (Tan *et al.*, 2017). This computational framework uses ideas from gravitation by considering the search agents as objects with mass. Thus, the gravitational forces dominate the motion of these masses - where lighter masses interact by gravitating towards heavier masses. The solutions are ranked based on the mass of the object. The higher the mass, the better the solution candidate. The gravitational forces hence act as a communication channel between the masses. The position of the masses correlate to the solution space in the objective space while the masses characterize the fitness landscape. As the iterations progress, and more gravitational interactions occur, it is expected that the masses would eventually swarm and aggregate to their fittest positions; providing the optimal solution to the problem.

The GSA algorithm randomly initializes a distribution of masses, $m_i(t)$ (search agents) in the objective space. Thus the initial positions for these masses, x_i^d are fixed. For a maximization problem, the most fit mass, $m^{best}(t)$ and the least fit mass, $m^{worst}(t)$ at time t are determined as follows:

$$m^{best}(t) = \max_{j \in [1,N]} m_j(t) \tag{5.19}$$

$$m^{worst}(t) = \min_{j \in [1,N]} m_j(t) \tag{5.20}$$

As for a minimization problem, the procedure is simply vice versa. The inertial mass, $m'_j(t)$ and gravitational masses, $M_j(t)$ are then computed based on the developed fitness map.

$$m'_i(t) = \frac{m_i(t) - m^{worst}(t)}{m^{best}(t) - m^{worst}(t)} \tag{5.21}$$

$$M_i(t) = \frac{m_i(t)}{\sum_{j=1}^{N} m_j(t)} \tag{5.22}$$

such that,

$$M_{ai} = M_{pi} = M_{ii} = M_i : i \in [1, N] \tag{5.23}$$

where M_{ai} and M_{pi} represent active and passive gravitational particle respectively. Consequently the gravitational constant, $G(t+1)$ and the Euclidean distance, $R_{ij}(t)$ is computed as the following:

$$G(t+1) = G(t)\exp\left(\frac{-\alpha t}{T_{\max}}\right) \tag{5.24}$$

$$R_{ij}(t) = \sqrt{\left(x_i(t)\right)^2 - \left(x_j(t)\right)^2} \tag{5.25}$$

where α is an arbitrary constant and T_{max} is the maximum number of iterations, $x_i(t)$ and $x_j(t)$ are the positions of particle i and j at time t. The interaction forces at time t, $F_{ij}^{d}(t)$ for each of the masses are then computed:

$$F_{ij}^d(t) = G(t)\left(\frac{M_{pi}(t) \times M_{aj}(t)}{R_{ij}(t) + \varepsilon}\right)\left(x_j^d(t) - x_i^d(t)\right) \tag{5.26}$$

where ε is a low-valued parameter. The sum of forces acting on each mass i is defined stochastically:

$$F_i^d(t) = \sum_{\substack{j=1 \\ i \neq j}}^{N} rand(w_j) F_{ij}^d(t) : rand(w_j) \in [0,1] \tag{5.27}$$

where $rand(w_j)$ is a randomly assigned weight. The magnitude of acceleration of each of the masses, $a_i^d(t)$ is then as follows:

$$a_i^d(t) = \left(\frac{F_i^d(t)}{M_{ii}(t)} \right) \tag{5.28}$$

Once the particle acceleration is determined, the particle positions and velocities are then computed:

$$v_i^d(t+1) = rand(w_j) + v_i^d(t) + a_i^d(t) \tag{5.29}$$

$$x_i^d(t+1) = x_i^d(t(t) + v_i^d(t(t+1) \tag{5.30}$$

where $rand(w_j)$ is a randomly assigned weight. The computational iterations are then continued until all the masses are at their most fit positions in the fitness landscape. The program is then halted when the stopping criterion is respected. The computational procedures of the GSA algorithm are presented in Algorithm 5.2 and the parameter settings are given in Table 2.

Algorithm 5.2: Gravitational Search Algorithm (GSA)

Step 1: Define no of particles, m_i, initial positions, $x_i(0)$ and define algorithm parameters $G(0)$, α.
Step 2: Randomly define the initial distribution of masses, $m(0)$ and/or update solutions
Step 3: Compute gravitational & inertial masses based on the

```
fitness map
Step 4: Compute the gravitational constant, G(t)
Step 5: Compute distance between agents, R_ij(t)
Step 6: Compute total force, F_i^d(t) and the acceleration a_i^d(t)
of each agent.
Step 7: Compute new velocity v_i(t) and position x_i(t) for each
agent
Step 8: If the fitness criterion is satisfied and t= T_max, halt
and print solutions
         else proceed to step 3
```

PARTICLE SWARM OPTIMZATION

PSO has been employed in many real-world problems ranging across various fields (Shi *et al.*, 2015; Pan and Das, 2016). PSO is based on the investigation of swarming or flocking behaviors among some species of organisms (such as; birds, fishes, etc.). In the spirit of evolutionary computing, PSO explores the objective space for optimal regions and then evaluates the found solutions with respect to a fitness condition. The optimal solutions obtained by this algorithm are achieved as a result of swarming particles moving through the objective space. The velocity and position updating rule plays a critical role in this technique. The velocities and the positions of each particle are updated as follows:

$$v_i(t+1) = wv_i(t) + c_1r_1[\hat{x}_i(t) - x_i(t)] + c_2r_2[g(t) - x_i(t)] \tag{5.31}$$

$$x_i(t+1) = x_i(t) + v_i(t+1) \tag{5.32}$$

Table 2. GSA settings

Parameters	Values
Initial parameter (G_o)	100
Number of mass agents, n	40
Constant parameter, α	20
Constant parameter, ε	0.01

where each particle is identified by the index *i*, $v_i(t)$ is the particle velocity and $x_i(t)$ is the particle position at iteration *t*. The expressions, $c_1 r_1[\hat{x}_i(t) - x_i(t)]$ and $c_2 r_2[g(t) - x_i(t)]$ represent the personal and social influence respectively. These terms influence the degree of exploration and exploitation of the PSO technique during the search. The parameters *w*, c_1 and c_2 are user- defined while r_1 and r_2 are randomly initialized. These parameters are typically constrained with the following ranges:

$$w \in [0,1.2],\ c_1 \in [0,2],\ c_2 \in [0,2],\ r_1 \in [0,1],\ r_2 \in [0,1] \quad (5.33)$$

The PSO program comes to a halt when all particles/candidate solutions have achieved their highest fitness during the iterations. The computational procedures of the PSO along with the initial parameters used are shown in Algorithm 5.3 and Table 3 respectively.

Algorithm 5.3: Particle Swarm Optimization (PSO) Algorithm

Step 1: Initialize no of particles, *i* and the algorithm parameters w, c_1, c_2, r_1, r_2, n_o
Step 2: Randomly initialize the positions $x_i(t)$ and velocities, $v_i(t)$
Step 3: Update solutions and/or compute individual and social influences
Step 4: Compute positions $x_i(t+1)$ and velocities $v_i(t+1)$ at next iteration
Step 5: If the swarm evolution time, $t > t_o + T_{max}$, update position x_i and velocity v_i and go to Step 3, else proceed to Step 6
Step 6: Evaluate fitness swarm.
Step 7: If fitness criterion is satisfied, halt program and print solutions, else go to step 3.

LÉVY - ENHANCED GRAVITATIONAL SEARCH ALGORITHM

Similar to the conventional GSA, the Lévy-enhanced GSA (Lévy-GSA) contains a stochastic generator. The stochastic generator of conventional metaheuristics function to randomly generate the initial population of agents. In conventional metaheuristics, the stochastic generator produces random variables based on a Gaussian distribution. For instance a random variable, $x \in X$ which is distributed with a mean, μ and variance, σ^2 is Gaussian or normally distributed when the probability distribution function is as follows:

$$G_X(x) = \frac{1}{\sigma\sqrt{2\pi}} \exp\left[-\frac{1}{2}\left(\frac{x-\mu}{\sigma}\right)^2\right] \tag{5.34}$$

The Gaussian distribution is widely applicable in various fields of studies for modelling real-valued random numbers - e.g. in studies that involve modeling Brownian motion (Sottinen and Tudor, 2006) and Monte Carlo simulations (Corney and Drummond, 2004). In this work, the standard normal distribution with $\mu = 0$ and $\sigma = 1$ is employed in the stochastic engine to generate random values in the conventional GSA and PSO techniques. On the other hand the CS algorithm (in its original form) does not use a Gaussian generator. CS employs Lévy flight (a heavy-tailed random walk probability distribution) to generate random numbers (see equations (5.17) and (5.18)). To further boost the performance of the GSA approach, the Lévy flight-based stochastic generator was integrated into the GSA approach. Other components of the GSA was maintained and not augmented. As with the conventional CS algorithm, the consequent iterations are randomized as well using Lévy flight. The algorithm and process flow for the Lévy-GSA is as in Algorithm 5.4 and Figure 1:

Table 3. PSO Settings

Parameters	Values
Initial parameter (c_1, c_2, w)	(1, 1.2, 0.8)
Number of particles	6
initial social influence (s_1, s_2, s_3, s_4, s_5, s_6)	(1.1, 1.05,1.033, 1.025, 1.02, 1.017)
initial personal influence (p_1, p_2, p_3, p_4, p_5, p_6)	(3, 4, 5, 6, 7,8)

Figure 1. Process flow of Lévy-GSA approach

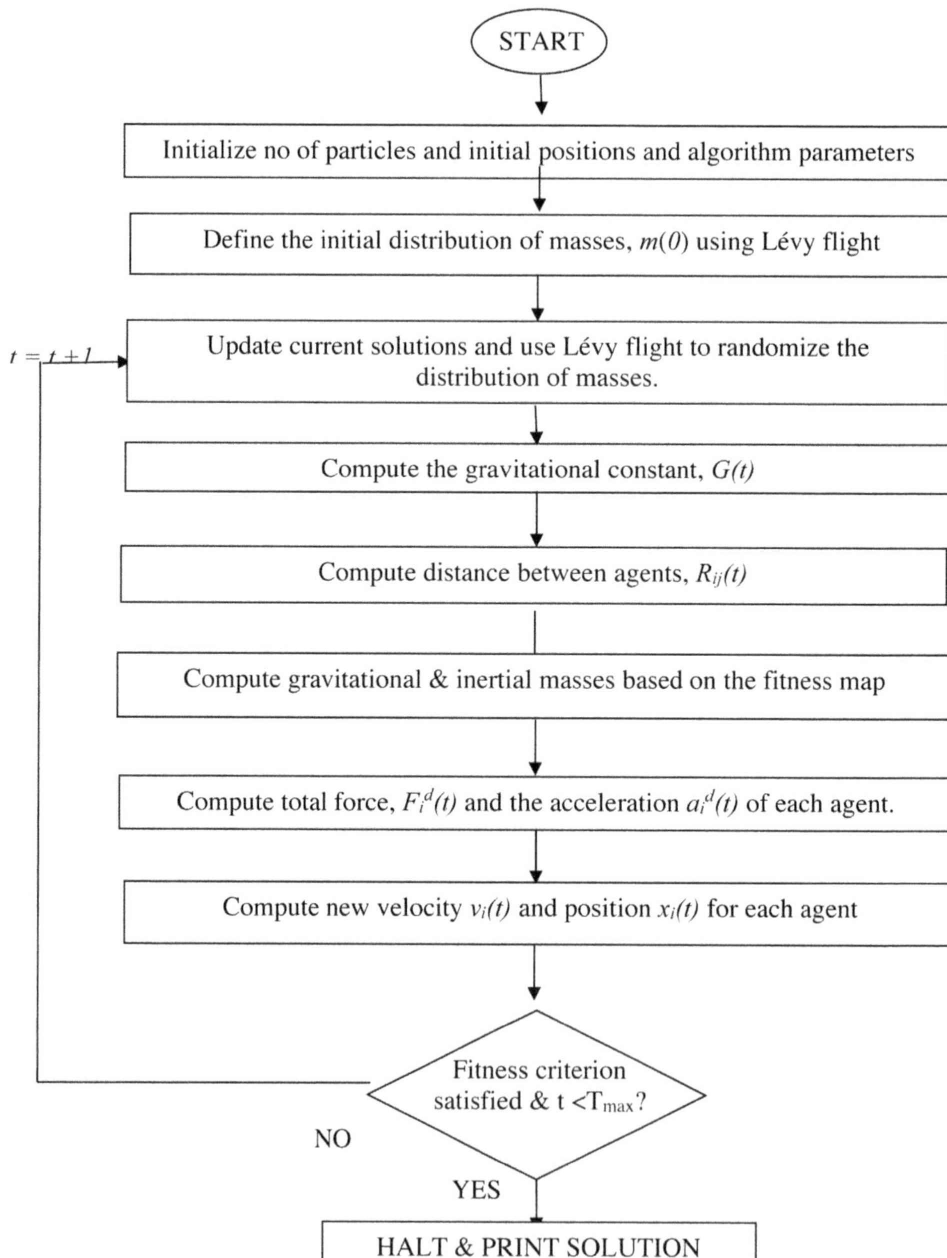

Algorithm 5.4: Lévy-GSA

Step 1: Define total particles, m_i, algorithm parameters, $G(0)$ & α and initial positions, $x_i(0)$
Step 2: Define the initial distribution of masses, $m(0)$ using Lévy flight/update current solutions and use Lévy flight to randomize the distribution of masses.
Step 3: Compute gravitational & inertial masses based on the fitness map
Step 4: Compute the gravitational constant, $G(t)$
Step 5: Compute distance between agents, $R_{ij}(t)$
Step 6: Compute total force, $F_i^d(t)$ and the acceleration $a_i^d(t)$ of each agent.
Step 7: Compute new velocity $v_i(t)$ and position $x_i(t)$ for each agent
Step 8: If the fitness criterion is satisfied and $t= T_{max}$, halt and print solutions
else proceed to step 2

ANALYSES

The weighted sum approach was employed to tackle the MO biofuel supply chain problem. All three objectives were aggregated to form a single weighted objective function - where all the weights were in the ranges of 0 to 1. Since the techniques employed in this work are stochastic by nature, the techniques were executed 3 times and the best solution was taken. The weights were varied into 28 values. Therefore each technique in this work was executed 84 times. The individual solutions depicting the Pareto frontier in this work were ranked into three classes; best, median and worst. The ranking was done based on the values obtained using the hypervolume indicator (HVI) (Jiang *et al.*, 2015; Bringmann and Friedrich, 2013). The HVI is a useful metric which measures dominance levels of solutions to a MO optimization problem. The ranked solutions for the CS approach is given in Table 4.

The Pareto frontier obtained using the CS technique is depicted in Figure 2.

As for GSA approach, the ranked individual solutions are given in Table 5 while its associated Pareto frontier is depicted in Figure 3.

Table 6 and Figure 4 provides the individual solution ranking and the Pareto frontier obtained using the PSO algorithm.

Table 4. Ranked individual solutions for the CS technique

Description		Best	Median	Worst
weights	w_1	0.3	0.2	0.6
	w_2	0.2	0.6	0.3
	w_3	0.5	0.2	0.1
Objective functions	*PR*	513,600,000	511,207,000	513,417,000
	SW	1,018,670	1,021,820	1,020,790
	Env	142,326	101,600	70,408.1
Iterations	*t*	52	44	49
Metric	*HVI*	7307.18	5193.35	3594.96

As mentioned in the previous section, the GSA was improved by modifying its stochastic generator - utilizing a Lévy flight-based mechanism instead of the conventional Gaussian random number generator. The ranked individual solutions and the Pareto frontier of the Lévy-GSA is given in Table 7 and figure 5.

Figure 2. Pareto frontier produced using the CS technique

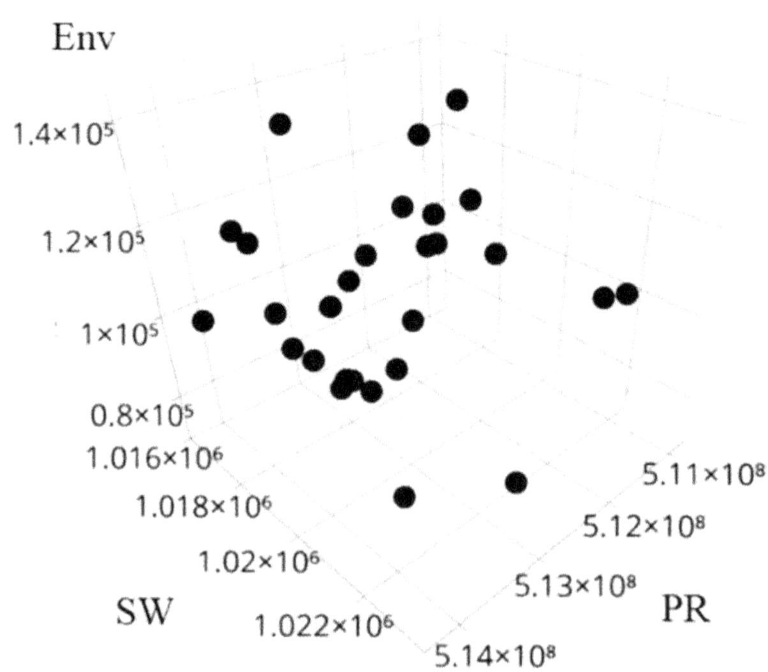

Table 5. Ranked individual solutions for the GSA technique

Description		Best	Median	Worst
weights	w_1	0.7	0.3	0.4
	w_2	0.2	0.4	0.3
	w_3	0.1	0.3	0.3
Objective functions	*PR*	208,859,000	155,942,000	124,922,000
	SW	1,004,340	1,004,780	1,005,040
	Env	334,224	378,267	351,498
Iterations	*t*	113	109	103
Metric	*HVI*	6887.16	5814.94	4321.9

In this work, the PSO, CS and GSA techniques were employed to solve the MO biofuel supply chain problem. The level of dominance for the entire Pareto frontier produced by each of the techniques were measured. Observing that the GSA technique produced the highest level of dominance, the mentioned

Figure 3. Pareto frontier produced using the GSA technique

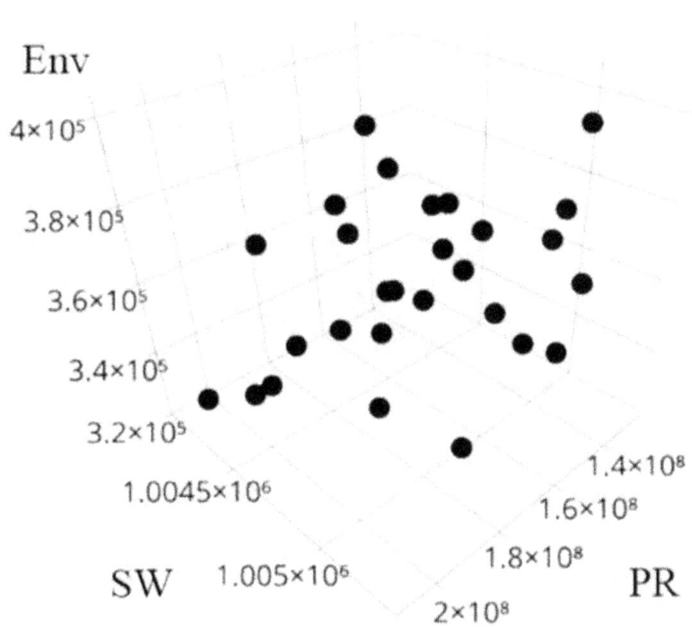

Table 6. Ranked individual solutions for the PSO technique

Description		Best	Median	Worst
weights	w_1	0.1	0.5	0.3
	w_2	0.2	0.2	0.6
	w_3	0.7	0.3	0.1
Objective functions	*PR*	215,783,000	127,843,000	91,770,000
	SW	1,003,790	1,004,500	1,004,770
	Env	313,382	400,193	377,767
Iterations	*t*	137	116	113
Metric	*HVI*	6667.77	5035.63	3402.03

technique was further improved (Lévy-GSA). The dominance values for the entire frontier produced by all the techniques employed in this work are shown in Figure 6.

Figure 4. Pareto frontier produced using the PSO technique

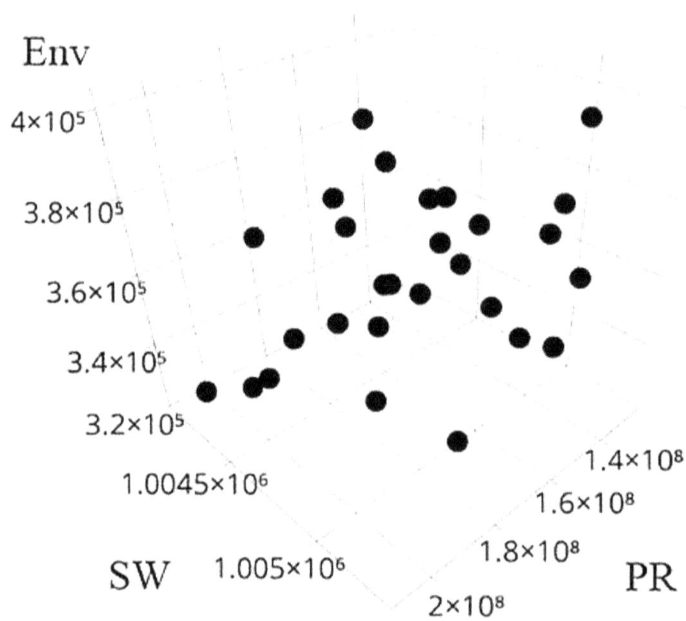

Table 7. Ranked individual solutions for the Lévy-GSA technique

Description		Best	Median	Worst
weights	w_1	0.2	0.2	0.4
	w_2	0.7	0.4	0.5
	w_3	0.1	0.4	0.1
Objective functions	*PR*	355,546,000	355,754,000	355,673,000
	SW	1,000,030	1,000,030	1,000,030
	Env	530,329	495,880	447,273
Iterations	*t*	300	300	300
Metric	*HVI*	18580.0395	17381.03197	15670.29223

In Figure 6, it can be seen that the Pareto frontier generated using the Lévy-GSA technique is significantly more dominant as compared all the other techniques. The frontier produced by Lévy-GSA technique outperforms all frontiers generated by other techniques. The Lévy-GSA performed almost twice as efficient as compared to all the other techniques employed in this work (in terms of frontier dominance). The Lévy-GSA technique boosted the

Figure 5. Pareto frontier produced using the Lévy-GSA technique

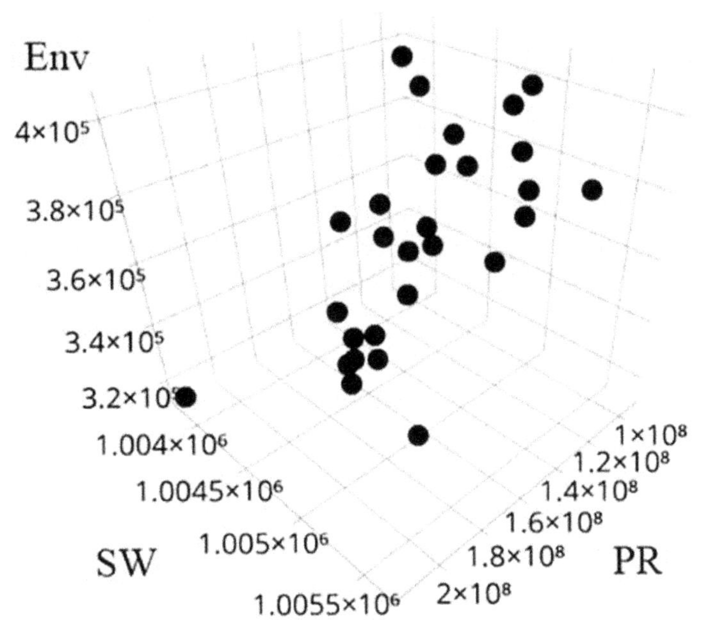

Figure 6. Dominance levels for Pareto frontiers generated by all four techniques

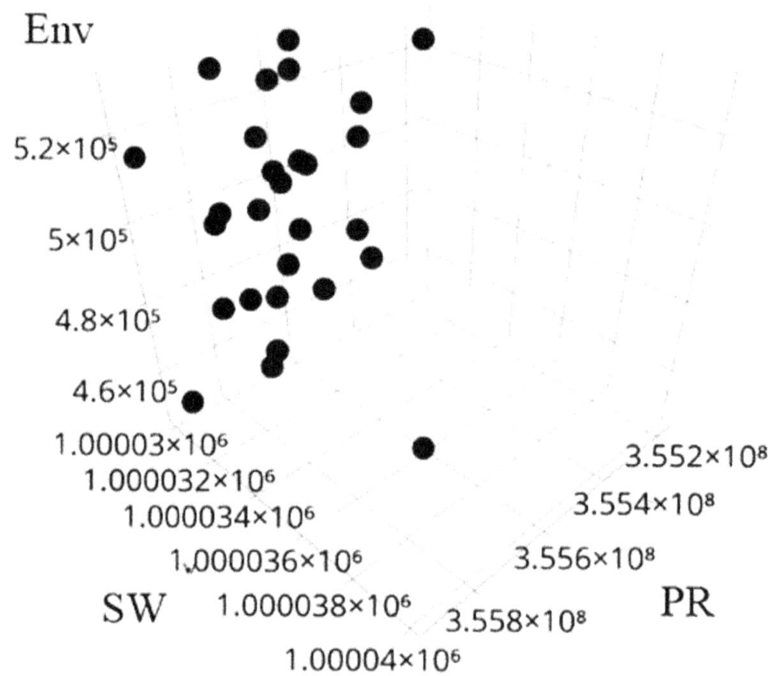

GSA technique by 200.03%, generating the most dominant Pareto frontier. This is followed by the frontiers generated by the GSA, CS and PSO techniques respectively.

In Figures 5.2 – 5.4, it can be observed that the CS, GSA and PSO approaches produce frontiers which are very diverse covering the objective space sparsely. Thus it is possible that many solutions were not concentrated in the optimal regions as compared to the Lévy-GSA technique. In the frontier generated by the Lévy-GSA approach (Figure 5.5) it is observed that the solutions seem more focused in certain regions of the objective space; which is the most optimal location.

The frontier dominance generated by the techniques also reflects the dominance levels of its best individual solution. This can be observed in Tables 5.4 – 5.6; where the best individual solutions produced by the CS, GSA and PSO techniques are 7307.18, 6887.16 and 6667.77. Similar to the dominance level of the entire frontier produced by the Lévy-GSA technique, the dominance of its individual solution far exceed those of the other approaches by a significantly large margin, 18,580.04 (see Table 7). The computational time (in seconds) taken by each technique employed in this work to construct the entire Pareto frontier is shown in Figure 7.

The Pareto frontier generated using the Lévy-GSA technique takes the longest computational time followed by the GSA, PSO and CS techniques respectively. This behavior is correlated to the dominance levels of the solutions generated by the techniques. Since the Lévy-GSA and GSA approaches produce the most dominant solutions, the region of the objective space explored is very wide. Thus, longer time taken to scrutinize the large number of solutions. On the contrary PSO and CS techniques take lesser time since the region of exploration is much smaller. Additionally it is also possible that these techniques often prematurely converge at sub-optimal regions in the objective space.

In terms of algorithmic complexity, the Lévy-GSA technique is not significantly more complex than other techniques utilized in this work. Besides the change in its stochastic generator, no additional computational steps has been added into the ones existing in the GSA technique. Thus the extended computational time may not have been caused by the increase in algorithmic complexity. In this work, all techniques produced converged solutions and no divergence issues were observed during program execution. All solutions produced were feasible; where no constraints were violated and the decision variables (solutions) obtained were non-negative. The fitness requirements defined in all the techniques were respected during program execution. All

Figure 7. Computational time taken by each technique to construct the entire Pareto frontier

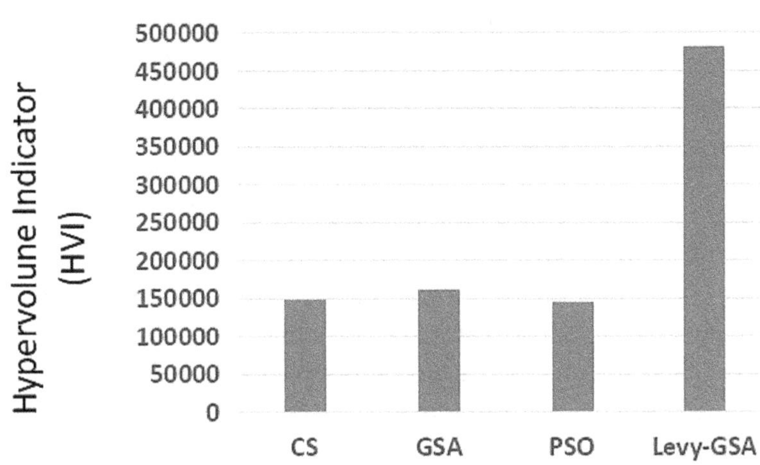

techniques employed in this work performed stable computations during program execution.

The Lévy-GSA approach outperformed all other swarm intelligence tehcniques employed in this work in terms of Pareto frontier dominance. This was followed by the frontiers produced by the GSA, CS and PSO techniques. The individual solutions constituting the Pareto frontier produced by the Lévy-GSA technique was observed to concentrate on the optimal region in the objective space (see Figure 5). With respect to the best individual solution, the CS approach outperforms the GSA and PSO techniques while the Lévy-GSA approach proves to be far superior than all the other algorithms (Tables 4-7).

FINAL REMARKS

The Lévy flight component was seen to improve the performance of the conventional GSA. In future works, the Lévy flight mechanism could be employed as a stochastic generator to other metaheuristics. As seen in this chapter, this type of modification to the original metaheuristic significantly contributes to the improvement of its performance. In addition, hybrid approaches could be employed to facilitate the integration of the Lévy flight component into the metaheuristic. The Normal Boundary Intersection (NBI) approach could also be employed to further enhance the quality of the solutions obtained (Ahmadi *et al.*, 2015).

REFERENCES

Abdelsalam, H. M., & Elassal, M. M. (2014). Joint economic lot sizing problem for a three—Layer supply chain with stochastic demand. *International Journal of Production Economics*, *155*, 272–283. doi:10.1016/j.ijpe.2014.01.015

Ahi, P., & Searcy, C. (2013). A comparative literature analysis of definitions for green and sustainable supply chain management. *Journal of Cleaner Production*, *52*, 329–341. doi:10.1016/j.jclepro.2013.02.018

Ahmadi, A., Kaymanesh, A., Siano, P., Janghorbani, M., Nezhad, A. E., & Sarno, D. (2015). Evaluating the effectiveness of normal boundary intersection method for short-term environmental/economic hydrothermal self-scheduling. *Electric Power Systems Research*, *123*, 192–204. doi:10.1016/j.epsr.2015.02.007

Akemann, G. (2017). Random matrix theory and quantum chromodynamics. *Stochastic Processes and Random Matrices: Lecture Notes of the Les Houches Summer School, 104*, 228.

Awudu, I., & Zhang, J. (2013). Stochastic production planning for a biofuel supply chain under demand and price uncertainties. *Applied Energy, 103*, 189–196. doi:10.1016/j.apenergy.2012.09.025

Ba, B. H., Prins, C., & Prodhon, C. (2016). Models for optimization and performance evaluation of biomass supply chains: An Operations Research perspective. *Renewable Energy, 87*, 977–989. doi:10.1016/j.renene.2015.07.045

Beenakker, C. W. J. (2015). Random-matrix theory of Majorana fermions and topological superconductors. *Reviews of Modern Physics, 87*(3), 1037–1066. doi:10.1103/RevModPhys.87.1037

Brandenburg, M., Govindan, K., Sarkis, J., & Seuring, S. (2014). Quantitative models for sustainable supply chain management: Developments and directions. *European Journal of Operational Research, 233*(2), 299–312. doi:10.1016/j.ejor.2013.09.032

Bringmann, K., & Friedrich, T. (2013). Approximation quality of the hypervolume indicator. *Artificial Intelligence, 195*, 265–290. doi:10.1016/j.artint.2012.09.005

Castillo-Villar, K. K. (2014). Metaheuristic algorithms applied to bioenergy supply chain problems: Theory, review, challenges, and future. *Energies, 7*(11), 7640–7672. doi:10.3390/en7117640

Chen, Y. W., Wang, L. C., Wang, A., & Chen, T. L. (2017). A particle swarm approach for optimizing a multi-stage closed loop supply chain for the solar cell industry. *Robotics and Computer-integrated Manufacturing, 43*, 111–123. doi:10.1016/j.rcim.2015.10.006

Corney, J. F., & Drummond, P. D. (2004). Gaussian Quantum Monte Carlo Methods for Fermions and Bosons'. *Physical Review Letters, 93*(26), 260401. doi:10.1103/PhysRevLett.93.260401 PMID:15697955

De Meyer, A., Cattrysse, D., Rasinmäki, J., & Van Orshoven, J. (2014). Methods to _ptimiza the design and management of biomass-for-bioenergy supply chains: A review. *Renewable & Sustainable Energy Reviews, 31*, 657–670. doi:10.1016/j.rser.2013.12.036

Divyesh, V.R., Sahu, C., Kirubakaran, V., Radhakrishnan, T.K., & Guruprasath, M. (2017). Energy optimization using metaheuristic bat algorithm assisted controller tuning for industrial and residential applications. *Transactions of the Institute of Measurement and Control*.

Ganesan, T., Elamvazuthi, I., & Vasant, P. (2016). Swarm Intelligence for Multiobjective Optimization of Extraction Process. In Handbook of Research on Modern Optimization Algorithms and Applications in Engineering and Economics (pp. 516-544). IGI Global. doi:10.4018/978-1-4666-9644-0.ch020

Gao, J., & You, F. (2015). Shale gas supply chain design and operations toward better economic and life cycle environmental performance: MINLP model and global optimization algorithm. *ACS Sustainable Chemistry & Engineering*, *3*(7), 1282–1291. doi:10.1021/acssuschemeng.5b00122

Gao, M., Ding, P., & Yang, Y. (2015, September). Time-Optimal Trajectory Planning of Industrial Robots Based on Particle Swarm Optimization. In *Instrumentation and Measurement, Computer, Communication and Control (IMCCC), 2015 Fifth International Conference on* (pp. 1934-1939). IEEE. 10.1109/IMCCC.2015.410

Ghaderi, H., Pishvaee, M. S., & Moini, A. (2016). Biomass supply chain network design: An optimization-oriented review and analysis. *Industrial Crops and Products*, *94*, 972–1000. doi:10.1016/j.indcrop.2016.09.027

Jiang, S., Zhang, J., Ong, Y. S., Zhang, A. N., & Tan, P. S. (2015). A simple and fast hypervolume indicator-based multiobjective evolutionary algorithm. *IEEE Transactions on Cybernetics*, *45*(10), 2202–2213. doi:10.1109/TCYB.2014.2367526 PMID:25474815

Joshi, A. S., Kulkarni, O., Kakandikar, G. M., & Nandedkar, V. M. (2017). Cuckoo Search Optimization-A Review. *Materials Today: Proceedings*, *4*(8), 7262–7269.

Kumar, R. S., Kondapaneni, K., Dixit, V., Goswami, A., Thakur, L. S., & Tiwari, M. K. (2016). Multi-objective modeling of production and pollution routing problem with time window: A self-learning particle swarm optimization approach. *Computers & Industrial Engineering*, *99*, 29–40. doi:10.1016/j.cie.2015.07.003

Lin, T., Rodríguez, L. F., Shastri, Y. N., Hansen, A. C., & Ting, K. C. (2014). Integrated strategic and tactical biomass–biofuel supply chain optimization. *Bioresource Technology*, *156*, 256–266. doi:10.1016/j.biortech.2013.12.121 PMID:24508904

Ling, S. H., Chan, K. Y., Leung, F. H. F., Jiang, F., & Nguyen, H. (2016). Quality and robustness improvement for real world industrial systems using a fuzzy particle swarm optimization. *Engineering Applications of Artificial Intelligence*, *47*, 68–80. doi:10.1016/j.engappai.2015.03.003

Mareli, M., & Twala, B. (2017). *An adaptive Cuckoo search algorithm for _ptimization*. Applied Computing and Informatics.

Marufuzzaman, M., Eksioglu, S. D., & Huang, Y. E. (2014). Two-stage stochastic programming supply chain model for biodiesel production via wastewater treatment. *Computers & Operations Research*, *49*, 1–17. doi:10.1016/j.cor.2014.03.010

Mastrocinque, E., Yuce, B., Lambiase, A., & Packianather, M. S. (2013). A multi-objective optimization for supply chain network using the bees algorithm. *International Journal of Engineering Business Management*, *5*, 38. doi:10.5772/56754

Mattos, C. L., Barreto, G. A., Horstkemper, D., & Hellingrath, B. (2017, June). Metaheuristic optimization for automatic clustering of customer-oriented supply chain data. In *Self-Organizing Maps and Learning Vector Quantization, Clustering and Data Visualization (WSOM), 2017 12th International Workshop on* (pp. 1-8). IEEE. 10.1109/WSOM.2017.8020025

Ogunbanwo, A., Williamson, A., Veluscek, M., Izsak, R., Kalganova, T., & Broomhead, P. (2014). Transportation network optimization. In *Encyclopedia of Business Analytics and Optimization* (pp. 2570–2583). IGI Global. doi:10.4018/978-1-4666-5202-6.ch229

Osmani, A., & Zhang, J. (2013). Stochastic optimization of a multi-feedstock lignocellulosic-based bioethanol supply chain under multiple uncertainties. *Energy*, *59*, 157–172. doi:10.1016/j.energy.2013.07.043

Osmani, A., & Zhang, J. (2014). Economic and environmental optimization of a large scale sustainable dual feedstock lignocellulosic-based bioethanol supply chain in a stochastic environment. *Applied Energy*, *114*, 572–587. doi:10.1016/j.apenergy.2013.10.024

Pan, I., & Das, S. (2016). Fractional order fuzzy control of hybrid power system with renewable generation using chaotic PSO. *ISA Transactions*, *62*, 19–29. doi:10.1016/j.isatra.2015.03.003 PMID:25816968

Poudel, S. R., Marufuzzaman, M., & Bian, L. (2016). Designing a reliable bio-fuel supply chain network considering link failure probabilities. *Computers & Industrial Engineering*, *91*, 85–99. doi:10.1016/j.cie.2015.11.002

Seuring, S. (2013). A review of modeling approaches for sustainable supply chain management. *Decision Support Systems*, *54*(4), 1513–1520. doi:10.1016/j.dss.2012.05.053

Shi, J., Zhang, W., Zhang, Y., Xue, F., & Yang, T. (2015). MPPT for PV systems based on a dormant PSO algorithm. *Electric Power Systems Research*, *123*, 100–107. doi:10.1016/j.epsr.2015.02.001

Sottinen, T., & Tudor, C. A. (2006). On the Equivalence of Multiparameter Gaussian Processes. *Journal of Theoretical Probability*, *19*(2), 461–485. doi:10.100710959-006-0022-5

Srivastav, A., & Agrawal, S. (2017). *Multi-objective optimization of slow moving inventory system using cuckoo search*. Intelligent Automation & Soft Computing.

Yeh, W. C., Lin, W. T., Lai, C. M., Lee, Y. C., Chung, Y. Y., & Lin, J. S. (2016, July). Application of simplified swarm optimization algorithm in deteriorate supply chain network problem. In *Evolutionary Computation (CEC), 2016 IEEE Congress on* (pp. 2695-2700). IEEE. 10.1109/CEC.2016.7744127

Yue, D., You, F., & Snyder, S. W. (2014). Biomass-to-bioenergy and biofuel supply chain optimization: Overview, key issues and challenges. *Computers & Chemical Engineering*, *66*, 36–56. doi:10.1016/j.compchemeng.2013.11.016

Zhang, J., Osmani, A., Awudu, I., & Gonela, V. (2013). An integrated optimization model for switchgrass-based bioethanol supply chain. *Applied Energy*, *102*, 1205–1217. doi:10.1016/j.apenergy.2012.06.054

APPENDIX

Nomenclature and Abbreviations

Biofuel Supply Chain Parameters

AC: abatement cost of carbon dioxide [yuan/kg]
$CEicb_{ik}$: increment of carbon dioxide emissions with loading each additional ton of biomass fuel per kilometer when broker k collects biomass fuel i [kg/t and km]
$AQmax_{i,t}$: maximum available quantity of local biomass fuel i in month t [t/month]
ACS: average electricity consumer surplus [yuan/kWh]
$CEitp_k$: increment of carbon dioxide emissions with loading each additional ton of fuel per kilometer when broker k transports biomass fuel to biomass power plant [kg/t and km]
$CEncb_{ik}$: carbon dioxide emissions per kilometer when broker k collects biomass fuel I with no-load conveyance [kg/km]
$CEntp_k$: carbon dioxide emissions per kilometer when broker k transports biomass fuel to biomass power plant with no load conveyance [kg/km]
CET: carbon dioxide emissions of thermal power plant for unit power generation [kg/kWh]
Dcb_{ik}: average transport distance when broker k collecting biomass fuel i [km]
Dtp_k: transport distance between broker k and biomass power plant [km]
E: efficiency of biomass power plant [decimal fraction]
EC: electricity consumption rate of biomass power plant [decimal fraction]
$extraY1$: first extra cost of excessive biomass power plant fuel inventory [yuan/month]
$extraY2$: second extra cost of excessive biomass power plant fuel inventory [yuan/month]
ER_k: expected return of broker k [decimal fraction mass/year]
FCb_k: fixed cost of broker k [yuan/year]
FCp: fixed cost of biomass power plant [yuan/year]
GC: unit generation cost of biomass power plant [yuan/kWh]
GS: government subsidies to biomass power generation [yuan/kWh]
GT: government tax revenues from biomass power plant [yuan/year]
HV_i: heat value of biomass fuel i [kJ/kg]
HVe: heat value of electricity [kJ/kWh]

HVmax: maximum heat value of mixed fuel [kJ/kg]
HV_{min}: minimum heat value of mixed fuel [kJ/kg]
$IQ_{i,0}$: inventory quantity of biomass fuel *i* at the beginning of month *1* [tonnes]
CEcb: carbon dioxide emissions during collecting biomass fuel [kg]
CEtp: carbon dioxide emissions during transporting biomass fuel to biomass power plant [kg]
EIC: extra inventory cost of biomass power plant [yuan]
$IQ_{i,t}$: inventory quantity of biomass fuel i at the end of month t [tonnes]
PP_i: purchase price of biomass fuel *i* from brokers [yuan/t]
$PQ_{ik,t}$: purchase quantity of biomass fuel *i* by broker *k* in month *t* [t]
q_t: electricity generation of biomass power plant in month t [kWh/month]
R_t: conversion rate from biomass fuel to electricity in month t [kg/kWh]
IQ_{max}: maximum inventory quantity of biomass power plant [t]
IL: rate of inventory loss [decimal fraction/month]
$LCcb_{ik}$: load capacity of conveyance when broker *k* collects biomass fuel *i* [t]
$LCtp_{ik}$: load capacity of conveyance when broker *k* transports biomass fuel *i* to biomass power plant [t]
$MCmax_i$: maximum moisture content of biomass fuel *i* required by biomass power plant [decimal fraction mass]
$MCorii_t$: original moisture content of biomass fuel *i* in month *t* [decimal fraction mass]
$MCaft_{ik}$: moisture content of biomass fuel *i* after processing by broker *k* [decimal fraction mass]
P: on-grid price of biomass power plant [yuan/kWh]
$PQ_{max, k}$: maximum purchasing quantity of biomass fuel by broker *k* [t/month]
q_{max}: maximum monthly electricity generation quantity of biomass power plant [kWh/month]
Q_{max}: maximum annual electricity generation quantity of biomass power plant [kWh/year]
Q_{min}: minimum annual electricity generation of biomass power plant [kWh/year]
$RIub_1$: first upper bound of reasonable fuel inventory [t]
$RIub_2$: second upper bound of reasonable fuel inventory [t]
$SIlb_i$: lower bound of safety inventory for biomass fuel *i* [t]
SC: unit storage cost of biomass power plant [yuan/month]
$SQmin_{ik}$: minimum supply quantity of biomass fuel *i* from broker
$TCcb_{ik}$: *a*verage unit transportation cost of broker *k* when collecting biomass fuel *I* [yuan/km]

$TCtp_{ik}$: average unit transportation cost of broker k when transporting biomass fuel i to biomass power plant [yuan/km]

$WR_{ik,t}$: ratio of the weight of biomass fuel i after processing to the weight before processing by broker k in month t [decimal fraction mass]

$AP_{ik,t}$: average price of broker k buying biomass fuel i in month t [yuan/t]

$BC_{i,t}$: biomass fuel i consumption in month t [t]

$BR_{i,t}$: blending ratio of biomass fuel i in mixed fuel in month t [decimal fraction mass]

CER: carbon dioxide emissions reduction [kg]

CET_{eq}: carbon dioxide emissions of thermal power plant for power generation equal to biomass power plant [kg]

CEB: carbon dioxide emissions of biomass power plant [kg]

$SQ_{ik,t}$: supply quantity of biomass fuel i by broker k in month t [t]

VCp: total variable cost of biomass power plant [yuan/year]

$Y1t$: binary variable to determine whether the inventory is over $RIub1$ at the end of month t

$Y2t$: binary variable to determine whether the inventory is over $RIub2$ at the end of month t

Chapter 6
Biofuel Supply Chain Optimization Using Random Matrix Generators

ABSTRACT

Supply chain problems are large-scale problems with complex interlinked variables. This sort of characteristic closely resembles structures often encountered in the nuclei of heavy atoms (e.g., platinum, gold or rhodium). Such structures are said to have the property of universality.

OVERVIEW

Ever since then, random matrix theory (RMT) has been used extensively for modeling large complex structures with highly interlinked components in various fields: solid state physics (Verbaarschot, 2004; Beenakker, 2015), quantum chromodynamics (Akemann, 2017), quantum information theory (Collins and Nechita, 2016), transport optimization (Krbálek and Seba, 2000), Big Data (Qiu and Antonik, 2017) and finance (Zhaoben *et al.*, 2014). Since supply chain networks share many key characteristics with some of the previously mentioned systems, these networks may naturally contain the property of universality. Therefore in this work, RMT is utilized to improve the conventional metaheuristic (CS) by modifying the stochastic generator component. Instead of employing the conventional Gaussian stochastic generator, this component was replaced with a RMT-based generator. The

DOI: 10.4018/978-1-7998-1710-9.ch006

resulting four RMT-based CS algorithms were implemented on the biofuel supply chain problem: Gaussian Unitary Ensemble CS (GUE-CS), Gaussian Orthogonal Ensemble CS (GOE-CS), Gaussian Symplectic Ensemble CS (GSE-CS) and Poisson-based CS (Poisson-CS).

RANDOM MATRIX THEORY

Random Matrix Theory (RMT) is a robust mathematical technique used to describe the behavior of complex systems. Due to its applicability to a vast range of systems, RMT is known to exhibit universality – a property of global symmetries shared by many systems within a certain symmetry class. In some sense RMT is a generalization of conventional statistical systems that often consists of matrices with fixed eigenvalues. In RMT, the eigenvalues themselves are statistically variant fitting into a probability distribution. This way RMT deals with the statistical behavior of the eigenvalues of random matrices. On the other hand, the random matrices have stochastic entries which are described by a distinct probability distribution function. Therefore in RMT there exists two probability distributions describing: the random matrix entries and the eigenvalue spread. The nearest neighbor spacing probability distribution (Schweiner *et al.*, 2017) of eigenvalues is given by Wigner's Surmise:

$$P(s) = A_i s^i \exp(-B_i s^2) \tag{6.1}$$

where s is the eigenvalue spacing, A_i and B_i are constant parameters. The normalized spacing, s and the mean spacing $\langle s \rangle$ is as follows:

$$s = \left[\frac{\lambda_{n+1} - \lambda_n}{\langle s \rangle} \right] \text{such that } \langle s \rangle = \langle \lambda_{n+1} - \lambda_n \rangle \tag{6.2}$$

where λ_n is the nth eigenvalue sequentially such that $\lambda_1 < ... < \lambda_n < \lambda_{n+1}$. The first type of random matrices are those that are modeled based on complex quantum systems (which have chaotic classical counterparts). The index i represent systems falling into this type of RMT theory; whereby $i = 1, 2$ and 4 denotes the Gaussian Orthogonal Ensemble (GOE) (Krishnan *et al.*, 2017),

Gaussian Unitary Ensemble (GUE) (Arguin *et al.*, 2017) and Gaussian Symplectic Ensemble (GSE) (Rehemanjiang *et al.*, 2016) spacing distributions respectively (Schierenberg *et al.*, 2012). These spacing distributions are given as follows:

1. 1. Gaussian Orthogonal Ensemble (GOE)

$$P_1(s) = \frac{\pi}{2} s \exp\left(-\frac{\pi}{4} s^2\right) \tag{6.3}$$

2. 2. Gaussian Unitary Ensemble (GUE)

$$P_2(s) = \frac{32}{\pi^2} s^2 \exp\left(-\frac{4}{\pi} s^2\right) \tag{6.4}$$

3. 3. Gaussian Symplectic Ensemble (GSE)

$$P_4(s) = \frac{2^{18}}{3^6 \pi^3} s^4 \exp\left(-\frac{64}{9\pi} s^2\right) \tag{6.5}$$

The second type of random matrix is employed to model complex quantum systems which have integrable classical counterpart (non-chaotic). The spacing distribution for such random matrices usually obey the Poisson process:

$$P_o(s) = \exp(-s) \tag{6.6}$$

These ensembles describe the probability density functions governing the random matrix entries. The constants, A_i and B_i are selected such that the following averaging properties are respected:

$$\int_0^\infty dsP_i\left(s\right) = 1 \tag{6.7}$$

$$\int_{0}^{\infty} dsP_i(s)s = 1 \tag{6.8}$$

RANDOM MATRIX GENERATORS

Conventional metaheuristics are equipped with stochastic generators also known as the random generator component of the algorithm. This component randomly initializes the search of the metaheuristic, positioning the start point of the search process in the objective space. The technique then iteratively improves the quality of the solution as the search progresses. Previous works have been done where different types of stochastic generators were experimented and their impacts on the performance of certain metaheuristics were evaluated (Ganesan *et al.*, 2018; Ganesan *et al.*, 2016a; Ganesan *et al.*, 2016b). In those works, it was observed that the changes in the type of stochastic generator influences the results of the optimization.

RMT could be considered as a generalization of conventional statistical systems. RMT was initially founded and used to solve problems involving systems with high levels of complexity - where the eigenstates of the systems are non-stationary and conventional stochastic techniques breakdown. Examples of such systems are: the behavior of heavy nuclei (Firk and Miller, 2009), energy flows in molecules (Leitner, 2005), stability of large ecosystems (Gibbs *et al.*, 2017) and properties of a gas of charged particles (Coulomb gas) (Cugliandolo, 2017).

In essence RMT deals with complex systems which consists of a network of many interacting components. Similar problems have been found in real-world optimization – where the problems are large-scale and often contain many interacting parameters which are correlated in a complicated way. The natural extension to conventional metaheuristics which enables it to deal with large-scale complex problems would be the generalization of its stochastic generator using RMT (creating a random matrix generator). In this spirit, the probability distribution function which randomizes the metaheuristics' initialization is modified using RMT. The proposed algorithmic technique for creating a random matrix generator is as follows:

Algorithm 6.1: Random Matrix Generator

```
Step 1: Generate random eigenvalue spacings, s from a Poisson/
GUE/GOE/GSE
Step 2: Determine the average eigenvalue spacing, Δλ
Step 3: Set initial eigenvalue, λ_0
Step 4: Set initial n x n matrix, H_ij
Step 5: Determine consequent eigenvalues, λ_i
```

$\lambda_{i+1} = \lambda_i + \Delta\lambda$

```
Step 6: Determine n x 1 eigenvector, E_i :
```

$E_i = \sum_j H_{ij} + \lambda_i$

```
Step 7: Generate random variables from a Gaussian probability
distribution function endowed eigenvector as the variance,
```

$\sigma^2 = E_i$:

$$P_i(x) = \frac{1}{\sqrt{2\pi E_i^2}} \exp\left(-\frac{(x-\mu)^2}{2E_i^2} \right)$$

ANALYSES

In this work, the objective functions of the MO biofuel supply chain problem was combined into a single function using the weighted sum approach. This procedure effectively transforms the MO problem into a single-objective optimization problem which can be solved for different weight values. The summing property for the weights are given as follows:

$$\sum_{i=1,..,m} w_i = 1 \text{ such that } w_i \in (0,1) \tag{6.9}$$

where the index i represents the individual weights and m is the maximum number of objectives (which is equivalent to the maximum number of weights). Due to the stochastic nature of the algorithms employed in this work, the algorithms were executed multiple times (3 executions) and the best solution was sampled. 28 variations of the weights were considered to construct the

Pareto frontier - at which the individual solutions were classified as best, worst and median. The individual solutions were measured and ranked using the hypervolume indicator (HVI) (Jiang *et al.*, 2015; Bringmann and Friedrich, 2013). The HVI effectively measures the dominance levels of solutions to a MO optimization problem. While using the HVI, the nadir point is commonly employed as a reference (or basis) during measurement. Hence the nadir point is considered as the most non-dominant point - where its exact value is used consistently on all the results obtained in this work. Considering the nadir point, the HVI is computed and scaled as follows:

$$HVI = \left[\frac{(x_1 - 10^6)(x_2 - 10^4)(x_3 - 10^3)}{10^{16}} \right] \tag{6.10}$$

where $x_{1,}$ x_2 and x_3 are individual candidate solutions. The ranked weighted individual solutions obtained using the CS approach is provided in Table 1. The entire frontier constructed using the CS approach is shown in Figure 1.

Using the Poisson-CS approach, the resulting ranked individual solutions is given in Table 2 while its associated Pareto frontier is depicted in Figure 2.

Table 3 and Figure 3 provides the ranked individual solutions and the Pareto frontier obtained using the GSE-CS algorithm.

The ranked individual solutions as well as their respective weights for the GOE-CS and GUE-CS are given in Tables 4 and 5, Figures 4 and 5 depicts the solutions representing the entire Pareto frontier.

Table 1. Ranked individual solutions for the CS technique

Description		Best	Median	Worst
weights	w_1	0.3	0.2	0.6
	w_2	0.2	0.6	0.3
	w_3	0.5	0.2	0.1
Objective functions	*PR*	513,600,000	511,207,000	513,417,000
	SW	1,018,670	1,021,820	1,020,790
	Env	142,326	101,600	70,408.1
Iterations	*t*	52	44	49
Metric	*HVI*	7,307.18	5,193.35	3,594.96

Figure 1. Pareto frontier from the CS technique

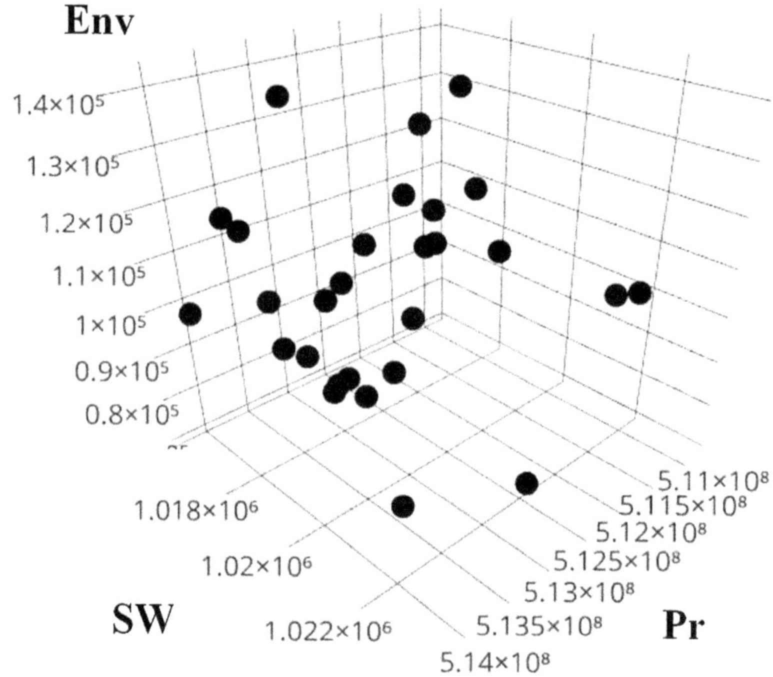

In addition, the HVI was used to measure the overall level of dominance for the entire Pareto frontier generated by each technique employed in this work. Figure 6 shows these degrees of dominance:

Figure 6 shows that the most dominant frontier was obtained using the GOE-CS followed by the conventional CS technique. Interestingly, the other

Table 2. Ranked individual solutions for the Poisson-CS technique

Description		Best	Median	Worst
weights	w_1	0.4	0.4	0.3
	w_2	0.4	0.5	0.4
	w_3	0.2	0.1	0.3
Objective functions	*PR*	245,476,000	442,101,000	282,918,000
	SW	1,006,110	1,004,330	1,004,940
	Env	462,609	6,895.43	1,117.03
Iterations	*t*	2	198	10
Metric	*HVI*	11,241.33264	258.5735346	3.282591965

Figure 2. Pareto frontier from the Poisson-CS technique

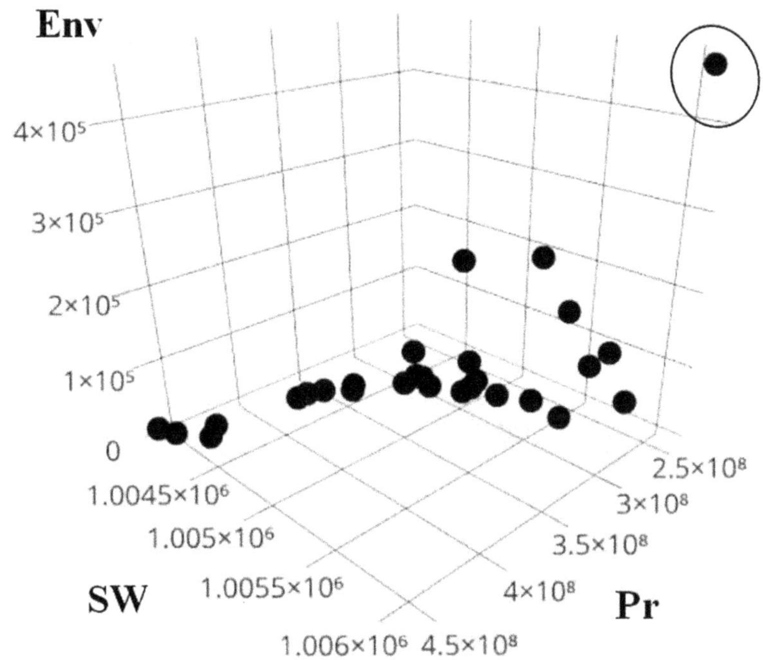

three techniques improved with RMT produced the least dominant frontiers - GUE-CS, Poisson-CS and GSE-CS respectively.

The most dominant individual solution was achieved by the GOE-CS technique followed by the Poisson-CS and CS techniques (see Tables 6.1, 6.2 and 6.4). The least dominant individual solutions were obtained using the GSE-CS and GUE-CS respectively (see Tables 6.3 and 6.5). To a high

Table 3. Ranked individual solutions for the GSE-CS technique

Description		Best	Median	Worst
weights	w_1	0.2	0.3	0.5
	w_2	0.5	0.6	0.2
	w_3	0.3	0.1	0.3
Objective functions	*PR*	457,778,000	307,741,000	353,240,000
	SW	1,004,270	1,004,690	1,004,450
	Env	36,541.3	2,741.15	1,050.19
Iterations	*t*	215	27	78
Metric	*HVI*	1,614.146034	53.12461162	1.758080756

Figure 3. Pareto frontier from the GSE-CS technique

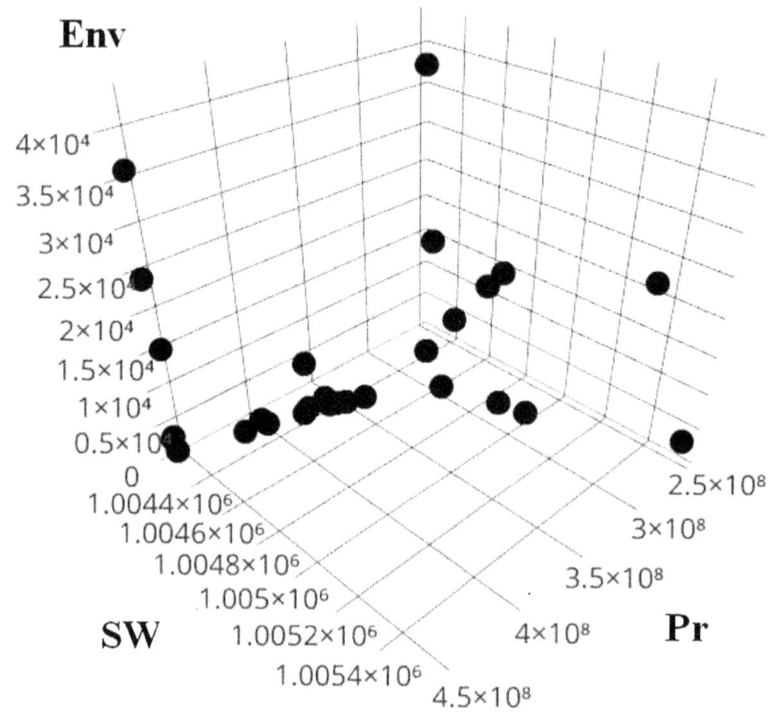

degree the dominance levels of the individual solutions were consistent with that of the entire Pareto frontier. For example in both instances the GOE-CS and CS techniques are the top three most dominant techniques. Similarly the GSE-CS, GUE-CS and the Poisson-CS came in lower in terms of dominance

Table 4. Ranked individual solutions for the GOE-CS technique

Description		Best	Median	Worst
weights	w_1	0.6	0.2	0.1
	w_2	0.2	0.2	0.7
	w_3	0.2	0.6	0.2
Objective functions	*PR*	320,196,000	311,556,000	314,295,000
	SW	1,004,690	1,004,800	1,004,820
	Env	405,964	200,519	141,623
Iterations	*t*	47	35	40
Metric	*HVI*	12,857.6503	6,163.962109	4,382.82702

Table 5. Ranked individual solutions for the GUE-CS technique

Description		Best	Median	Worst
weights	w_1	0.5	0.4	0.2
	w_2	0.3	0.2	0.7
	w_3	0.2	0.4	0.1
Objective functions	*PR*	337,985,000	365,898,000	304,589,000
	SW	1,004,570	1,004,440	1,004,660
	Env	156,623	17,765	1,158.42
Iterations	*t*	64	95	25
Metric	*HVI*	5,215.785325	608.3501587	4.783774438

in both respects. It could be observed that the techniques that managed to focus their individual solutions at the optimal region of the objective space (as indicated in Figure 4) have a higher overall level of dominance. Besides GOE-CS, the conventional CS technique generates many individual solutions at that region (refer to Figure 1). Other techniques tend to generate solutions

Figure 4. Pareto frontier from the GOE-CS technique

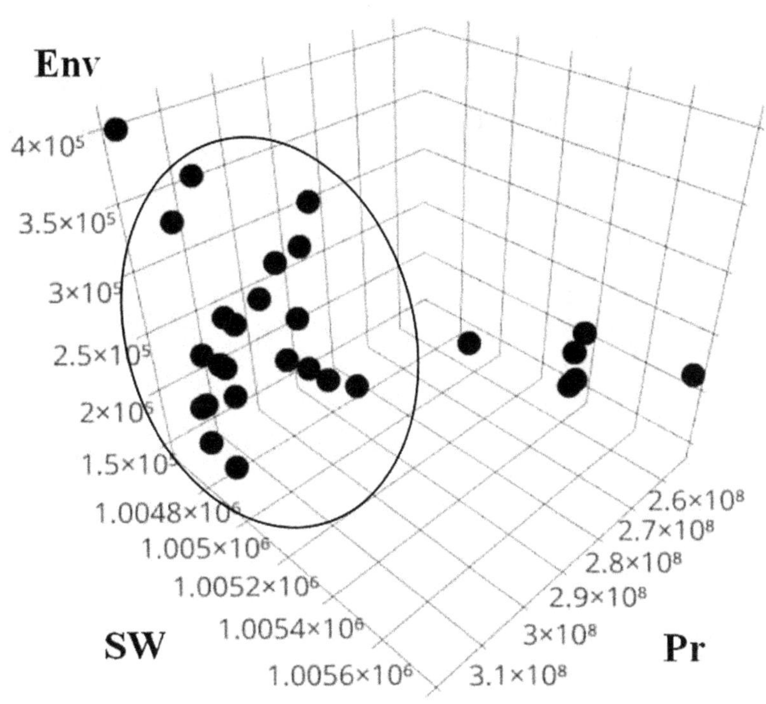

Figure 5. Pareto frontier from the GUE-CS technique

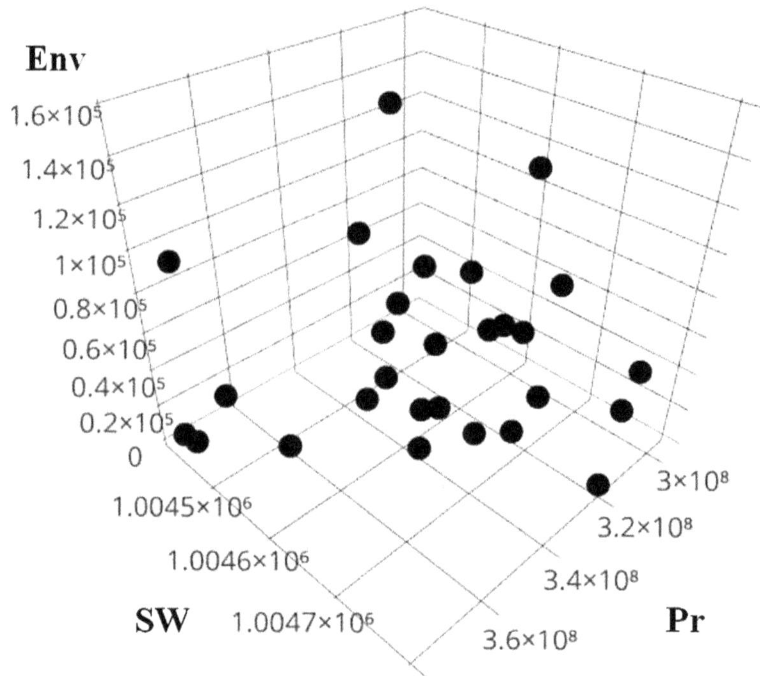

at other sub-optimal regions in the objective space. For instance the Poisson-CS technique (in Figure 2) generates solutions at very low values of the *Env*, objective function while the GSE-CS technique (in Figure 3) generates

Figure 6. Degrees of dominance of the all the Pareto frontiers

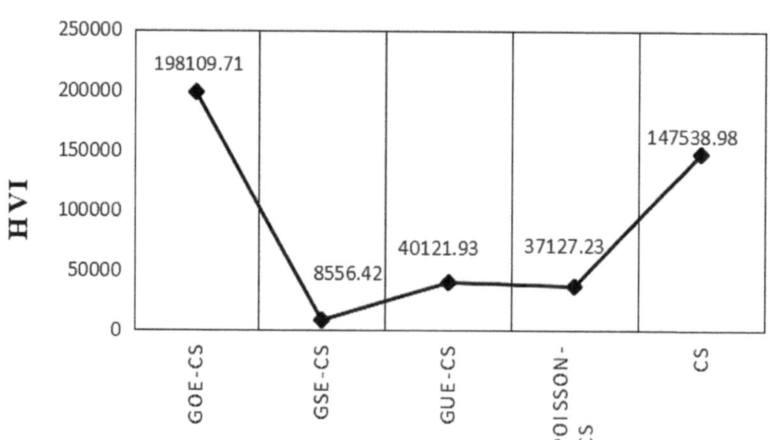

scattered individual solutions which are non-uniformly distributed. On the other hand, the GUE-CS (in Figure 5) produces a relatively uniform distribution of solutions in the objective space - however these solutions are not located in the optimal region.

It is important to note that the Poisson-CS seems to behave in a contradictory way when the dominance of its individual solutions are compared with the dominance of the entire Pareto frontier. The Poisson-CS can be seen to have a high level of individual solution dominance (second to the GOE-CS; 11,241.33) but an extremely weak frontier dominance; ranking second last to the GSE-CS approach (see Figure 6). It is possible that the Poisson-CS approach seems to be highly incompatible with the problem at hand in the 'No Free Lunch Theorem' sense (Whitley, 2014). Such incompatibility could be observed in the anomalous solutions generated by the technique (as indicated in Figure 2).

It may be that in this work, one of the risks of modifying the stochastic generator has been identified. Since the stochastic generator is a core component in metaheuristics, modifying it changes very fundamental mechanisms that make up an algorithm. This could significantly boost the technique's performance or it could negatively impact the technique by worsening its optimization performance - as observed in the case of the Poisson-CS technique in this work. The overall computational time taken for all the techniques utilized in this work to generate the entire Pareto frontier is shown in Figure 7.

The execution times of all the techniques in Figure 7 are observed to be consistent with the frontier dominance values (see Figure 6). The techniques generating the most dominant frontiers (GOE-CS, CS and GSE-CS) spent less time searching through the objective space as compared to the Poisson-CS as

Figure 7. Degrees of dominance of all the all the Pareto frontiers

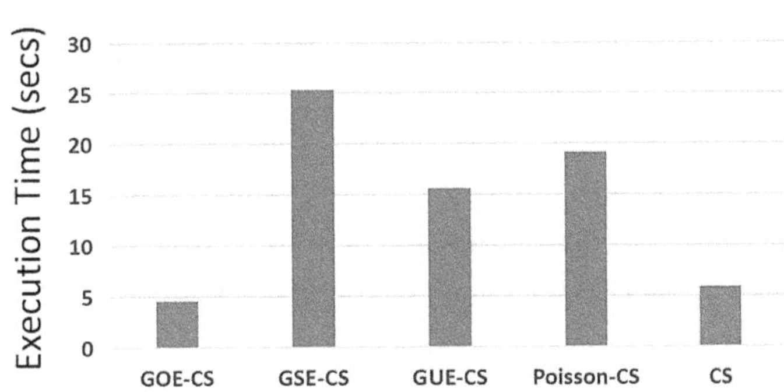

well as GUE-CS approaches. Due to their efficiency in reaching the optimal region in the objective space, execution time for these techniques are short.

In this work, the solutions produced by the all the techniques were found to be feasible as none of the constraints were violated. The algorithmic complexity of the techniques employed are fairly similar since no additional computational steps were added into the existing CS approach. The only difference is that the stochastic generators are replaced with different types of RMT-based distributions (Poisson, GUE, GOE and GSE). No premature convergence was observed during the execution of the techniques in this work. The fitness criteria specified in all the techniques were respected during program execution. All techniques implemented throughout this work performed stable computations.

In this work, the GOE-CS technique outperformed all other approaches. From the analysis of the results of the computational experiments, it can be observed that the RMT-based stochastic generators could significantly affect the optimization performance of the CS method. The overall Pareto dominance produced by the GOE-CS was 34.276% higher than the conventional CS. The improved performance of the GOE-CS is reflected in its computational efficiency (20.931% faster convergence). Thus it can be seen that the RMT-based improvement to the conventional CS approach is very suitable to be employed in highly complex environments such as the MO biofuel supply chain problem.

OUTLOOK

In this chapter, although the GOE-CS performed better than the conventional CS, some RMT-based techniques was also observed to negatively impact algorithmic performance; as observed in the results produced by the Poisson-CS approach. Therefore further tests involving RMT-based stochastic generators could be conducted in future works. For instance, numerical experiments on the performance of RMT-based stochastic generators on other metaheuristics could be carried out - e.g. particle swarm optimization (Mousavi *et al.*, 2016), differential evolution (Ganesan *et al.*, 2015), bat algorithm (Singh *et al*, 2015) and fish swarm algorithm (Luan *et al.*, 2016). In addition, research could also be done on methods to counter the effects of the unsatisfactory performance encountered by some RMT-based techniques (like the Poisson-CS variant).

REFERENCES

Che, Z. (2017). Universality of random matrices with correlated entries. *Electronic Journal of Probability*, 22.

Collins, B., & Nechita, I. (2016). Random matrix techniques in quantum information theory. *Journal of Mathematical Physics*, *57*(1), 015215. doi:10.1063/1.4936880

Ganesan, T., Elamvazuthi, I., & Vasant, P. (2015). Multiobjective design optimization of a nano-CMOS voltage-controlled oscillator using game theoretic-differential evolution. *Applied Soft Computing*, *32*, 293–299. doi:10.1016/j.asoc.2015.03.016

Ganesan, T., Elamvazuthi, I., & Vasant, P. (2016b). Swarm Intelligence for Multiobjective Optimization of Extraction Process. In Handbook of Research on Modern Optimization Algorithms and Applications in Engineering and Economics (pp. 516-544). IGI Global. doi:10.4018/978-1-4666-9644-0.ch020

Ganesan, T., Vasant, P., & Elamvazuthi, I. (2016a). *Advances in Metaheuristics: Applications in Engineering Systems*. CRC Press. doi:10.1201/9781315297651

Krbálek, M., & Seba, P. (2000). The statistical properties of the city transport in Cuernavaca (Mexico) and random matrix ensembles. *Journal of Physics. A, Mathematical and General*, *33*(26), L229–L234. doi:10.1088/0305-4470/33/26/102

Luan, X. Y., Li, Z. P., & Liu, T. Z. (2016). A novel attribute reduction algorithm based on rough set and improved artificial fish swarm algorithm. *Neurocomputing*, *174*, 522–529. doi:10.1016/j.neucom.2015.06.090

Mousavi, S. M., Sadeghi, J., Niaki, S. T. A., & Tavana, M. (2016). A bi-objective inventory optimization model under inflation and discount using tuned Pareto-based algorithms: NSGA-II, NRGA, and MOPSO. *Applied Soft Computing*, *43*, 57–72. doi:10.1016/j.asoc.2016.02.014

Qiu, R. C., & Antonik, P. (2017). *Smart grid using big data analytics: a random matrix theory approach*. John Wiley & Sons. doi:10.1002/9781118716779

Singh, K., Vasant, P., Elamvazuthi, I., & Kannan, R. (2015). PID tuning of servo motor using bat algorithm. *Procedia Computer Science*, *60*, 1798–1808. doi:10.1016/j.procs.2015.08.290

Verbaarschot, J. (2004, December). The supersymmetric method in random matrix theory and applications to QCD. In AIP Conference Proceedings (Vol. 744, No. 1, pp. 277-362). AIP. doi:10.1063/1.1853204

Whitley, D. (2014). Sharpened and focused no free lunch and complexity theory. In *Search Methodologies* (pp. 451–476). Boston, MA: Springer. doi:10.1007/978-1-4614-6940-7_16

Zhaoben, F., Ying-chang, L., & Zhidong, B. (2014). *Spectral Theory of Large Dimensional Random Matrices and Its Applications to Wireless Communications and Finance Statistics: Random Matrix Theory and Its Applications*. World Scientific.

APPENDIX

Nomenclature and Abbreviations

Biofuel Supply Chain Parameters

AC: abatement cost of carbon dioxide [yuan/kg]
$CEicb_{ik}$: increment of carbon dioxide emissions with loading each additional ton of biomass fuel per kilometer when broker k collects biomass fuel i [kg/t and km]
$AQmax_{i,t}$: maximum available quantity of local biomass fuel i in month t [t/month]
ACS: average electricity consumer surplus [yuan/kWh]
$CEitp_{k}$: increment of carbon dioxide emissions with loading each additional ton of fuel per kilometer when broker k transports biomass fuel to biomass power plant [kg/t and km]
$CEncb_{ik}$: carbon dioxide emissions per kilometer when broker k collects biomass fuel I with no-load conveyance [kg/km]
$CEntp_{k}$: carbon dioxide emissions per kilometer when broker k transports biomass fuel to biomass power plant with no load conveyance [kg/km]
CET: carbon dioxide emissions of thermal power plant for unit power generation [kg/kWh]
Dcb_{ik}: average transport distance when broker k collecting biomass fuel i [km]
Dtp_{k}: transport distance between broker k and biomass power plant [km]
E: efficiency of biomass power plant [decimal fraction]

EC: electricity consumption rate of biomass power plant [decimal fraction]
extraY1: first extra cost of excessive biomass power plant fuel inventory [yuan/month]
extraY2: second extra cost of excessive biomass power plant fuel inventory [yuan/month]
ER_{k}: expected return of broker k [decimal fraction mass/year]
FCb_{k}: fixed cost of broker k [yuan/year]
FCp: fixed cost of biomass power plant [yuan/year]
GC: unit generation cost of biomass power plant [yuan/kWh]

GS: government subsidies to biomass power generation [yuan/kWh]
GT: government tax revenues from biomass power plant [yuan/year]
HV_i: heat value of biomass fuel *i* [kJ/kg]
HVe: heat value of electricity [kJ/kWh]
*HV*max: maximum heat value of mixed fuel [kJ/kg]
HV_{min}: minimum heat value of mixed fuel [kJ/kg]
$IQ_{i,0}$: inventory quantity of biomass fuel *i* at the beginning of month *1* [tonnes]
CEcb: carbon dioxide emissions during collecting biomass fuel [kg]
CEtp: carbon dioxide emissions during transporting biomass fuel to biomass power plant [kg]
EIC: extra inventory cost of biomass power plant [yuan]
$IQ_{i,t}$: inventory quantity of biomass fuel i at the end of month t [tonnes]
PP_i: purchase price of biomass fuel *i* from brokers [yuan/t]
$PQ_{ik,t}$: purchase quantity of biomass fuel *i* by broker *k* in month *t* [t]
q_t: electricity generation of biomass power plant in month t [kWh/month]
R_t: conversion rate from biomass fuel to electricity in month t [kg/kWh]
IQ_{max}: maximum inventory quantity of biomass power plant [t]
IL: rate of inventory loss [decimal fraction/month]
$LCcb_{ik}$: load capacity of conveyance when broker *k* collects biomass fuel *i* [t]
$LCtp_{ik}$: load capacity of conveyance when broker *k* transports biomass fuel *i* to biomass power plant [t]
$MCmax_i$: maximum moisture content of biomass fuel *i* required by biomass power plant [decimal fraction mass]
$MCorii_t$: original moisture content of biomass fuel *i* in month *t* [decimal fraction mass]
$MCaft_{ik}$: moisture content of biomass fuel *i* after processing by broker *k* [decimal fraction mass]
P: on-grid price of biomass power plant [yuan/kWh]
$PQ_{max,k}$: maximum purchasing quantity of biomass fuel by broker *k* [t/month]
q_{max}: maximum monthly electricity generation quantity of biomass power plant [kWh/month]
Q_{max}: maximum annual electricity generation quantity of biomass power plant [kWh/year]
Q_{min}: minimum annual electricity generation of biomass power plant [kWh/year]
$RIub_1$: first upper bound of reasonable fuel inventory [t]
$RIub_2$: second upper bound of reasonable fuel inventory [t]
$SIlb_i$: lower bound of safety inventory for biomass fuel *i* [t]
SC: unit storage cost of biomass power plant [yuan/month]

$SQmin_{ik}$: minimum supply quantity of biomass fuel *i* from broker

$TCcb_{ik}$: average unit transportation cost of broker *k* when collecting biomass fuel *I* [yuan/km]

$TCtp_{ik}$: average unit transportation cost of broker *k* when transporting biomass fuel *i* to biomass power plant [yuan/km]

$WR_{ik,t}$: ratio of the weight of biomass fuel *i* after processing to the weight before processing by broker *k* in month *t* [decimal fraction mass]

$AP_{ik,t}$: average price of broker *k* buying biomass fuel *i* in month *t* [yuan/t]

$BC_{i,t}$: biomass fuel *i* consumption in month *t* [t]

$BR_{i,t}$: blending ratio of biomass fuel *i* in mixed fuel in month *t* [decimal fraction mass]

CER: carbon dioxide emissions reduction [kg]

CET_{eq}: carbon dioxide emissions of thermal power plant for power generation equal to biomass power plant [kg]

CEB: carbon dioxide emissions of biomass power plant [kg]

$SQ_{ik,t}$: supply quantity of biomass fuel *i* by broker *k* in month *t* [t]

VCp: total variable cost of biomass power plant [yuan/year]

Y1t: binary variable to determine whether the inventory is over *RIub1* at the end of month *t*

Y2t: binary variable to determine whether the inventory is over *RIub2* at the end of month *t*

Conclusion

TOWARDS EFFICIENT ALGORITHMS FOR ENERGY OPTIMIZATION

The global energy industry has grown and transformed significantly in terms of complexity. Many aspects of the energy industry has undergone rapid advances - instigating decision makers to speedily adapt as they strive to remain relevant. Besides adaptation, some key players in the industry have also capitalized on the current energy landscape to give them a competitive edge against others. These organizations have taken an offensive stance in the energy market; acting as initiators of change instead of being mere followers. They have actively contributed to technological disruption by opening up new energy markets which challenges established products and practices. Such changes have influenced the very way decision makers think and thus understand the energy industry.

Computational intelligence (CI) techniques have played a pivotal role in tackling these challenges in the current energy industry. These techniques have aided the industrial practitioner when dealing with problems ranging across various aspects of the market: power plant operations/management, supply chain planning/ scheduling, power transmission/distribution, plant design and process engineering. Recent works involving these aspects of the industry was detailed in Chapter 1. There the computational techniques were seen to be broadly applied for predictive learning and optimization. These applications focused on the main global energy industries: renewable energy systems, distributed generation, nuclear plants, coal and gas power plants.

Real-world problems are often mathematically messy and plagued with various uncertainties. These issues become amplified when the problem is exposed to various complexities. Since the applications in this book were based on real-world situations, the problems often came in such a form. The three primary complexities faced by the industrial applications presented in this book are:

- High levels of uncertainty
- Multiple nonlinear target objectives
- Large-scale and interlinked parameters

High levels of uncertainty could be seen in Part 2 of Chapter 2, where the MO solar-powered irrigation system was optimized while factoring uncertainties in weather conditions (insolation and ambient temperature). Since the level of uncertainties in the data was high, the type-2 fuzzy logic framework was employed for modeling the system. To benchmark the performance of the metaheuristics, various metrics were employed: HVI, sigma diversity metric and the value of the aggregate objective function.

Besides uncertainties, the number of objectives in the application problem could significantly contribute to the complexity of the problem. For instance in Chapter 4, the GT-AC system consisted of four target objectives. Adding to the complexity, the four objective functions were all nonlinear. The metaheuristic employed (DE) had to be considerably enhanced using extreme value stochastic engines to deal with the problem at hand.

The MO biofuel supply chain problem given in Chapters 5 and 6 has a large-scale form where its parameters are close interconnected. Since the form of the problem closely resembles physical systems such as the nuclei of heavy atoms, the solution method was designed with mathematical techniques commonly employed to tackle such problems in nuclear physics; random matrix theory (RMT). Implementing the cuckoo search (CS) technique which was improved via RMT, the supply chain problem was solved successfully.

As the energy industry moves into a new era, practitioners face completely novel challenges in their day to day business operations. In these rapidly changing environments, metaheuristic techniques are very useful tools to have. The author hopes that the ideas and insights provided in this book would come in useful for engineers and decision makers in this new energy industry.

Related Readings

To continue IGI Global's long-standing tradition of advancing innovation through emerging research, please find below a compiled list of recommended IGI Global book chapters and journal articles in the areas of green cities, environmental management, and sustainable urban development. These related readings will provide additional information and guidance to further enrich your knowledge and assist you with your own research.

Ahuja, K., & Khosla, A. (2019). A Framework to Develop a Zero-Carbon Emission Sustainable Cognitive City. In K. Ahuja & A. Khosla (Eds.), *Driving the Development, Management, and Sustainability of Cognitive Cities* (pp. 1–26). Hershey, PA: IGI Global. doi:10.4018/978-1-5225-8085-0.ch001

Akkucuk, U. (2016). SCOR Model and the Green Supply Chain. In U. Akkucuk (Ed.), *Handbook of Research on Waste Management Techniques for Sustainability* (pp. 108–124). Hershey, PA: IGI Global. doi:10.4018/978-1-4666-9723-2.ch006

Albuquerque, C. M. (2017). Cities Really Smart and Inclusive: Possibilities and Limits for Social Inclusion and Participation. In L. Carvalho (Ed.), *Handbook of Research on Entrepreneurial Development and Innovation Within Smart Cities* (pp. 229–247). Hershey, PA: IGI Global. doi:10.4018/978-1-5225-1978-2.ch011

Angelucci, F., Di Girolamo, C., & Zazzero, E. (2018). New Designing Codes for Urban Infrastructures: A Hypothesis of a Transdisciplinary Approach. In G. Carlone, N. Martinelli, & F. Rotondo (Eds.), *Designing Grid Cities for Optimized Urban Development and Planning* (pp. 209–237). Hershey, PA: IGI Global. doi:10.4018/978-1-5225-3613-0.ch012

Avdimiotis, S., & Tilikidou, I. (2017). Smart Tourism Development: The Case of Halkidiki. In L. Carvalho (Ed.), *Handbook of Research on Entrepreneurial Development and Innovation Within Smart Cities* (pp. 491–513). Hershey, PA: IGI Global. doi:10.4018/978-1-5225-1978-2.ch021

Ayvaz, B., & Görener, A. (2016). Reverse Logistics in the Electronics Waste Industry. In U. Akkucuk (Ed.), *Handbook of Research on Waste Management Techniques for Sustainability* (pp. 155–171). Hershey, PA: IGI Global. doi:10.4018/978-1-4666-9723-2.ch008

Bagnato, V. P. (2018). Technology and Urban Structure: The Grid City Between Technological Innovation and New Public Space System. In G. Carlone, N. Martinelli, & F. Rotondo (Eds.), *Designing Grid Cities for Optimized Urban Development and Planning* (pp. 238–253). Hershey, PA: IGI Global. doi:10.4018/978-1-5225-3613-0.ch013

Bala, I., & Singh, G. (2019). Green Communication for Cognitive Cities. In K. Ahuja & A. Khosla (Eds.), *Driving the Development, Management, and Sustainability of Cognitive Cities* (pp. 87–110). Hershey, PA: IGI Global. doi:10.4018/978-1-5225-8085-0.ch004

Bernardino, S., & Santos, J. F. (2017). Building Smarter Cities through Social Entrepreneurship. In L. Carvalho (Ed.), *Handbook of Research on Entrepreneurial Development and Innovation Within Smart Cities* (pp. 327–362). Hershey, PA: IGI Global. doi:10.4018/978-1-5225-1978-2.ch015

Bernardo, M. D. (2017). Smart City Governance: From E-Government to Smart Governance. In L. Carvalho (Ed.), *Handbook of Research on Entrepreneurial Development and Innovation Within Smart Cities* (pp. 290–326). Hershey, PA: IGI Global. doi:10.4018/978-1-5225-1978-2.ch014

Bhatia, P., & Singh, P. (2019). Technological and Gamified Solutions for Pollution Control in Cognitive Cities. In K. Ahuja & A. Khosla (Eds.), *Driving the Development, Management, and Sustainability of Cognitive Cities* (pp. 234–249). Hershey, PA: IGI Global. doi:10.4018/978-1-5225-8085-0.ch010

Bhatt, R. (2017). Zero Tillage for Mitigating Global Warming Consequences and Improving Livelihoods in South Asia. In W. Ganpat & W. Isaac (Eds.), *Environmental Sustainability and Climate Change Adaptation Strategies* (pp. 126–161). Hershey, PA: IGI Global. doi:10.4018/978-1-5225-1607-1.ch005

Bolívar, M. P. (2018). Governance in Smart Cities: A Comparison of Practitioners' Perceptions and Prior Research. *International Journal of E-Planning Research*, *7*(2), 1–19. doi:10.4018/IJEPR.2018040101

Bostanci, S. H., & Albayrak, A. N. (2017). The Role of Eco-Municipalities in Climate Change for a Sustainable Future. In W. Ganpat & W. Isaac (Eds.), *Environmental Sustainability and Climate Change Adaptation Strategies* (pp. 213–231). Hershey, PA: IGI Global. doi:10.4018/978-1-5225-1607-1.ch008

Brkljačić, T., Majetić, F., & Tarabić, B. N. (2017). Smart Environment: Cyber Parks (Connecting Nature and Technology). In L. Carvalho (Ed.), *Handbook of Research on Entrepreneurial Development and Innovation Within Smart Cities* (pp. 150–172). Hershey, PA: IGI Global. doi:10.4018/978-1-5225-1978-2.ch008

Cai, Y. Z., Wu, F., Li, J., Wang, J., & Huang, M. (2019). Application of UAV Technology to Planning Study on Chinese Villages in Guanzhong. In J. Vargas-Hernández & J. Zdunek-Wielgołaska (Eds.), *Bioeconomical Solutions and Investments in Sustainable City Development* (pp. 180–195). Hershey, PA: IGI Global. doi:10.4018/978-1-5225-7958-8.ch008

Calisto, M. D., & Gonçalves, A. (2017). Smart Citizens, Wise Decisions: Sustainability-Driven Tourism Entrepreneurs. In L. Carvalho (Ed.), *Handbook of Research on Entrepreneurial Development and Innovation Within Smart Cities* (pp. 20–43). Hershey, PA: IGI Global. doi:10.4018/978-1-5225-1978-2.ch002

Capelo, C. (2017). Exploring the Dynamics of an Energy Service Venture. In L. Carvalho (Ed.), *Handbook of Research on Entrepreneurial Development and Innovation Within Smart Cities* (pp. 269–289). Hershey, PA: IGI Global. doi:10.4018/978-1-5225-1978-2.ch013

Caridi, G. (2018). Calabria 1783: The Orthogonal Grid as a Physical and Ideological Device of Reconstruction. In G. Carlone, N. Martinelli, & F. Rotondo (Eds.), *Designing Grid Cities for Optimized Urban Development and Planning* (pp. 176–187). Hershey, PA: IGI Global. doi:10.4018/978-1-5225-3613-0.ch010

Carlone, G. (2018). Cities and Extension Plans in the Kingdom of the Two Sicilies: Borgo Murattiano of Bari (1812-1859). In G. Carlone, N. Martinelli, & F. Rotondo (Eds.), *Designing Grid Cities for Optimized Urban Development and Planning* (pp. 1–18). Hershey, PA: IGI Global. doi:10.4018/978-1-5225-3613-0.ch001

Carvalho, L. C. (2017). Entrepreneurial Ecosystems: Lisbon as a Smart Start-Up City. In L. Carvalho (Ed.), *Handbook of Research on Entrepreneurial Development and Innovation Within Smart Cities* (pp. 1–19). Hershey, PA: IGI Global. doi:10.4018/978-1-5225-1978-2.ch001

Castagnolo, V. (2018). Analyzing, Classifying, Safeguarding: Drawing for the Borgo Murattiano Neighbourhood of Bari. In G. Carlone, N. Martinelli, & F. Rotondo (Eds.), *Designing Grid Cities for Optimized Urban Development and Planning* (pp. 93–108). Hershey, PA: IGI Global. doi:10.4018/978-1-5225-3613-0.ch006

Castilla-Polo, F., Gallardo-Vázquez, D., Sánchez-Hernández, M. I., & Ruiz-Rodríguez, M. D. (2017). Cooperatives as Responsible and Innovative Entrepreneurial Ecosystems in Smart Territories: The Olive Oil Industry in the South of Spain. In L. Carvalho (Ed.), *Handbook of Research on Entrepreneurial Development and Innovation Within Smart Cities* (pp. 459–490). Hershey, PA: IGI Global. doi:10.4018/978-1-5225-1978-2.ch020

Charalabidis, Y., & Theocharopoulou, C. (2019). A Participative Method for Prioritizing Smart City Interventions in Medium-Sized Municipalities. *International Journal of Public Administration in the Digital Age*, *6*(1), 41–63. doi:10.4018/IJPADA.2019010103

Chatsiwa, J., Mujere, N., & Maiyana, A. B. (2016). Municipal Solid Waste Management. In U. Akkucuk (Ed.), *Handbook of Research on Waste Management Techniques for Sustainability* (pp. 19–43). Hershey, PA: IGI Global. doi:10.4018/978-1-4666-9723-2.ch002

Chawla, R., Singhal, P., & Garg, A. K. (2019). Impact of Dust for Solar PV in Indian Scenario: Experimental Analysis. In K. Ahuja & A. Khosla (Eds.), *Driving the Development, Management, and Sustainability of Cognitive Cities* (pp. 111–138). Hershey, PA: IGI Global. doi:10.4018/978-1-5225-8085-0.ch005

Cılız, N., Yıldırım, H., & Temizel, Ş. (2016). Structure Development for Effective Medical Waste and Hazardous Waste Management System. In U. Akkucuk (Ed.), *Handbook of Research on Waste Management Techniques for Sustainability* (pp. 303–327). Hershey, PA: IGI Global. doi:10.4018/978-1-4666-9723-2.ch016

Collignon de Alba, C., Haberleithner, J., & López, M. M. (2017). Creative Industries in the Smart City: Overview of a Liability in Emerging Economies. In L. Carvalho (Ed.), *Handbook of Research on Entrepreneurial Development and Innovation Within Smart Cities* (pp. 107–126). Hershey, PA: IGI Global. doi:10.4018/978-1-5225-1978-2.ch006

Corum, A. (2016). Remanufacturing, an Added Value Product Recovery Strategy. In U. Akkucuk (Ed.), *Handbook of Research on Waste Management Techniques for Sustainability* (pp. 347–367). Hershey, PA: IGI Global. doi:10.4018/978-1-4666-9723-2.ch018

Crosas, C. (2018). Latin American Cities: Modern Grids From 1850s. In G. Carlone, N. Martinelli, & F. Rotondo (Eds.), *Designing Grid Cities for Optimized Urban Development and Planning* (pp. 39–51). Hershey, PA: IGI Global. doi:10.4018/978-1-5225-3613-0.ch003

D'Onofrio, S., Habenstein, A., & Portmann, E. (2019). Ontological Design for Cognitive Cities: The New Principle for Future Urban Management. In K. Ahuja & A. Khosla (Eds.), *Driving the Development, Management, and Sustainability of Cognitive Cities* (pp. 183–211). Hershey, PA: IGI Global. doi:10.4018/978-1-5225-8085-0.ch008

Dalzero, S. (2018). The Time of the Finished World Has Begun: A New Map of the World – National Borders Partially or Fully Fenced-Off. In G. Carlone, N. Martinelli, & F. Rotondo (Eds.), *Designing Grid Cities for Optimized Urban Development and Planning* (pp. 254–275). Hershey, PA: IGI Global. doi:10.4018/978-1-5225-3613-0.ch014

Damurski, L. (2016). Recent Progress in Online Communication Tools for Urban Planning: A Comparative Study of Polish and German Municipalities. *International Journal of E-Planning Research, 5*(1), 39–54. doi:10.4018/IJEPR.2016010103

Damurski, L. (2016). Smart City, Integrated Planning, and Multilevel Governance: A Conceptual Framework for e-Planning in Europe. *International Journal of E-Planning Research, 5*(4), 41–53. doi:10.4018/IJEPR.2016100103

Das, K. K., & Sharma, N. K. (2016). Post Disaster Housing Management for Sustainable Urban Development: A Review. *International Journal of Geotechnical Earthquake Engineering*, *7*(1), 1–18. doi:10.4018/IJGEE.2016010101

Das, S., & Nayyar, A. (2019). Innovative Ideas to Manage Urban Traffic Congestion in Cognitive Cities. In K. Ahuja & A. Khosla (Eds.), *Driving the Development, Management, and Sustainability of Cognitive Cities* (pp. 139–162). Hershey, PA: IGI Global. doi:10.4018/978-1-5225-8085-0.ch006

Deakin, M., & Reid, A. (2017). The Embedded Intelligence of Smart Cities: Urban Life, Citizenship, and Community. *International Journal of Public Administration in the Digital Age*, *4*(4), 62–74. doi:10.4018/IJPADA.2017100105

Djukic, A., Stupar, A., & Antonic, B. M. (2018). The Orthogonal Urban Matrix of the Towns in Vojvodina, Northern Serbia: Genesis and Transformation. In G. Carlone, N. Martinelli, & F. Rotondo (Eds.), *Designing Grid Cities for Optimized Urban Development and Planning* (pp. 128–156). Hershey, PA: IGI Global. doi:10.4018/978-1-5225-3613-0.ch008

Dolunay, O. (2016). A Paradigm Shift: Empowering Farmers to Eliminate the Waste in the Form of Fresh Water and Energy through the Implementation of 4R+T. In U. Akkucuk (Ed.), *Handbook of Research on Waste Management Techniques for Sustainability* (pp. 368–379). Hershey, PA: IGI Global. doi:10.4018/978-1-4666-9723-2.ch019

Dowling, C. M., Walsh, S. D., Purcell, S. M., Hynes, W. M., & Rhodes, M. L. (2017). Operationalising Sustainability within Smart Cities: Towards an Online Sustainability Indicator Tool. *International Journal of E-Planning Research*, *6*(4), 1–17. doi:10.4018/IJEPR.2017100101

Ekman, U. (2018). Smart City Planning: Complexity. *International Journal of E-Planning Research*, *7*(3), 1–21. doi:10.4018/IJEPR.2018070101

Erdogan, S. (2016). The Effect of Working Capital Management on Firm's Profitability: Evidence from Istanbul Stock Exchange. In U. Akkucuk (Ed.), *Handbook of Research on Waste Management Techniques for Sustainability* (pp. 244–261). Hershey, PA: IGI Global. doi:10.4018/978-1-4666-9723-2.ch013

Eudoxie, G., & Roopnarine, R. (2017). Climate Change Adaptation and Disaster Risk Management in the Caribbean. In W. Ganpat & W. Isaac (Eds.), *Environmental Sustainability and Climate Change Adaptation Strategies* (pp. 97–125). Hershey, PA: IGI Global. doi:10.4018/978-1-5225-1607-1.ch004

Fasolino, I. (2018). Rules for a New Town After a Disaster: The Gridded Schemes in the Plans. In G. Carlone, N. Martinelli, & F. Rotondo (Eds.), *Designing Grid Cities for Optimized Urban Development and Planning* (pp. 157–175). Hershey, PA: IGI Global. doi:10.4018/978-1-5225-3613-0.ch009

Fernandes, C. M., & Dias de Sousa, I. (2017). Digital Swarms: Social Interaction and Emergent Phenomena in Personal Communications Networks. In L. Carvalho (Ed.), *Handbook of Research on Entrepreneurial Development and Innovation Within Smart Cities* (pp. 44–59). Hershey, PA: IGI Global. doi:10.4018/978-1-5225-1978-2.ch003

Fernandes, V., Moreira, A., & Daniel, A. I. (2017). A Qualitative Analysis of Social Entrepreneurship Involving Social Innovation and Intervention. In L. Carvalho (Ed.), *Handbook of Research on Entrepreneurial Development and Innovation Within Smart Cities* (pp. 417–438). Hershey, PA: IGI Global. doi:10.4018/978-1-5225-1978-2.ch018

Gabriel, B. F., Valente, R. A., Dias-de-Oliveira, J., Neto, V. F., & Andrade-Campos, A. (2017). Methodologies for Engineering Learning and Teaching (MELT): An Overview of Engineering Education in Europe and a Novel Concept for Young Students. In L. Carvalho (Ed.), *Handbook of Research on Entrepreneurial Development and Innovation Within Smart Cities* (pp. 363–391). Hershey, PA: IGI Global. doi:10.4018/978-1-5225-1978-2.ch016

Garrick, T. A., & Liburd, O. E. (2017). Impact of Climate Change on a Key Agricultural Pest: Thrips. In W. Ganpat & W. Isaac (Eds.), *Environmental Sustainability and Climate Change Adaptation Strategies* (pp. 232–254). Hershey, PA: IGI Global. doi:10.4018/978-1-5225-1607-1.ch009

Gencer, Y. G. (2016). Mystery of Recycling: Glass and Aluminum Examples. In U. Akkucuk (Ed.), *Handbook of Research on Waste Management Techniques for Sustainability* (pp. 172–191). Hershey, PA: IGI Global. doi:10.4018/978-1-4666-9723-2.ch009

Gencer, Y. G., & Akkucuk, U. (2016). Reverse Logistics: Automobile Recalls and Other Conditions. In U. Akkucuk (Ed.), *Handbook of Research on Waste Management Techniques for Sustainability* (pp. 125–154). Hershey, PA: IGI Global. doi:10.4018/978-1-4666-9723-2.ch007

Gonçalves, J. M., Martins, T. G., & Vilhena da Cunha, I. B. (2017). Local Creative Ecosystems as a Strategy for the Development of Low-Density Urban Spaces. In L. Carvalho (Ed.), *Handbook of Research on Entrepreneurial Development and Innovation Within Smart Cities* (pp. 127–149). Hershey, PA: IGI Global. doi:10.4018/978-1-5225-1978-2.ch007

Goundar, S., & Appana, S. (2017). Mainstreaming Development Policies for Climate Change in Fiji: A Policy Gap Analysis and the Role of ICTs. In W. Ganpat & W. Isaac (Eds.), *Environmental Sustainability and Climate Change Adaptation Strategies* (pp. 1–31). Hershey, PA: IGI Global. doi:10.4018/978-1-5225-1607-1.ch001

Grochulska-Salak, M. (2019). Urban Farming in Sustainable City Development. In J. Vargas-Hernández & J. Zdunek-Wielgołaska (Eds.), *Bioeconomical Solutions and Investments in Sustainable City Development* (pp. 43–64). Hershey, PA: IGI Global. doi:10.4018/978-1-5225-7958-8.ch003

Gruszecka, K. (2019). Ecological Centre of Warsaw as a Development Path. In J. Vargas-Hernández & J. Zdunek-Wielgołaska (Eds.), *Bioeconomical Solutions and Investments in Sustainable City Development* (pp. 117–150). Hershey, PA: IGI Global. doi:10.4018/978-1-5225-7958-8.ch006

Hesapci-Sanaktekin, O., & Aslanbay, Y. (2016). The Networked Self: Collectivism Redefined in Civic Engagements through Social Media Causes. In U. Akkucuk (Ed.), *Handbook of Research on Waste Management Techniques for Sustainability* (pp. 262–276). Hershey, PA: IGI Global. doi:10.4018/978-1-4666-9723-2.ch014

Ikeda, M. (2019). Developing a Sustainable Eco-City in Pre-Olympic Tokyo: Potential of New Methods and Their Limits in an Urban Era. In J. Vargas-Hernández & J. Zdunek-Wielgołaska (Eds.), *Bioeconomical Solutions and Investments in Sustainable City Development* (pp. 196–223). Hershey, PA: IGI Global. doi:10.4018/978-1-5225-7958-8.ch009

İşcan, E. (2019). Strategies of Sustainable Bioeconomy in the Industry 4.0 Framework for Inclusive and Social Prosperity. In J. Vargas-Hernández & J. Zdunek-Wielgołaska (Eds.), *Bioeconomical Solutions and Investments in Sustainable City Development* (pp. 21–42). Hershey, PA: IGI Global. doi:10.4018/978-1-5225-7958-8.ch002

J., J., Samui, P., & Dixon, B. (2016). Determination of Rate of Medical Waste Generation Using RVM, MARS and MPMR. In U. Akkucuk (Ed.), *Handbook of Research on Waste Management Techniques for Sustainability* (pp. 1-18). Hershey, PA: IGI Global. doi:10.4018/978-1-4666-9723-2.ch001

Jiménez, R., Rodríguez, P. L., & Fernández, R. R. (2019). Green Spaces of the Metropolitan Area of Guadalajara. In J. Vargas-Hernández & J. Zdunek-Wielgołaska (Eds.), *Bioeconomical Solutions and Investments in Sustainable City Development* (pp. 151–179). Hershey, PA: IGI Global. doi:10.4018/978-1-5225-7958-8.ch007

Kais, S. M. (2017). Climate Change: Vulnerability and Resilience in Commercial Shrimp Aquaculture in Bangladesh. In W. Ganpat & W. Isaac (Eds.), *Environmental Sustainability and Climate Change Adaptation Strategies* (pp. 162–187). Hershey, PA: IGI Global. doi:10.4018/978-1-5225-1607-1.ch006

Khan, B. (2019). Bio-Economy: Visions, Strategies, and Policies. In J. Vargas-Hernández & J. Zdunek-Wielgołaska (Eds.), *Bioeconomical Solutions and Investments in Sustainable City Development* (pp. 1–20). Hershey, PA: IGI Global. doi:10.4018/978-1-5225-7958-8.ch001

Khanna, B. K. (2017). Indian National Strategy for Climate Change Adaptation and Mitigation. In W. Ganpat & W. Isaac (Eds.), *Environmental Sustainability and Climate Change Adaptation Strategies* (pp. 32–63). Hershey, PA: IGI Global. doi:10.4018/978-1-5225-1607-1.ch002

Khanna, B. K. (2017). Vulnerability of the Lakshadweep Coral Islands in India and Strategies for Mitigating Climate Change Impacts. In W. Ganpat & W. Isaac (Eds.), *Environmental Sustainability and Climate Change Adaptation Strategies* (pp. 64–96). Hershey, PA: IGI Global. doi:10.4018/978-1-5225-1607-1.ch003

Kocasoy, G. (2016). Economic Instruments for Sustainable Environmental Management. In U. Akkucuk (Ed.), *Handbook of Research on Waste Management Techniques for Sustainability* (pp. 192–211). Hershey, PA: IGI Global. doi:10.4018/978-1-4666-9723-2.ch010

Lallo, C. H., Smalling, S., Facey, A., & Hughes, M. (2017). The Impact of Climate Change on Small Ruminant Performance in Caribbean Communities. In W. Ganpat & W. Isaac (Eds.), *Environmental Sustainability and Climate Change Adaptation Strategies* (pp. 296–321). Hershey, PA: IGI Global. doi:10.4018/978-1-5225-1607-1.ch011

Li, Z., Wang, Y., & Chen, Q. (2017). Real-Time Monitoring of Intercity Passenger Flows Based on Big Data: A Decision Support Tool for Urban Sustainability. *International Journal of Strategic Decision Sciences*, *8*(4), 120–128. doi:10.4018/IJSDS.2017100106

Loi, N. K., Huyen, N. T., Tu, L. H., Tram, V. N., Liem, N. D., Dat, N. L., ... Minh, D. N. (2017). Sustainable Land Use and Watershed Management in Response to Climate Change Impacts: Case Study in Srepok Watershed, Central Highland of Vietnam. In W. Ganpat & W. Isaac (Eds.), *Environmental Sustainability and Climate Change Adaptation Strategies* (pp. 255–295). Hershey, PA: IGI Global. doi:10.4018/978-1-5225-1607-1.ch010

López-Arranz, M. A. (2017). The Role Corporate Social Responsibility Has in the Smart City Project in Spain. In L. Carvalho (Ed.), *Handbook of Research on Entrepreneurial Development and Innovation Within Smart Cities* (pp. 439–458). Hershey, PA: IGI Global. doi:10.4018/978-1-5225-1978-2.ch019

Lucas, M. R., Rego, C., Vieira, C., & Vieira, I. (2017). Proximity and Cooperation for Innovative Regional Development: The Case of the Science and Technology Park of Alentejo. In L. Carvalho (Ed.), *Handbook of Research on Entrepreneurial Development and Innovation Within Smart Cities* (pp. 199–228). Hershey, PA: IGI Global. doi:10.4018/978-1-5225-1978-2.ch010

Magrinho, A., Neves, J., & Silva, J. R. (2017). The Triple Helix Model: Evidence in the Internationalization of the Health Industry. In L. Carvalho (Ed.), *Handbook of Research on Entrepreneurial Development and Innovation Within Smart Cities* (pp. 60–79). Hershey, PA: IGI Global. doi:10.4018/978-1-5225-1978-2.ch004

Maiorano, A. C. (2018). Urban Fronts in Murattiano Neighbourhood of Bari: A Selective Survey of the Built Environment. In G. Carlone, N. Martinelli, & F. Rotondo (Eds.), *Designing Grid Cities for Optimized Urban Development and Planning* (pp. 78–92). Hershey, PA: IGI Global. doi:10.4018/978-1-5225-3613-0.ch005

Martinelli, N., & Mangialardi, G. (2018). Cities With Grid Layout: Ubiquitousness and Flexibility of an Urban Model. In G. Carlone, N. Martinelli, & F. Rotondo (Eds.), *Designing Grid Cities for Optimized Urban Development and Planning* (pp. 188–208). Hershey, PA: IGI Global. doi:10.4018/978-1-5225-3613-0.ch011

Mizutani, S., Liao, K., & Sasaki, T. G. (2019). Forest-River-Ocean Nexus-Based Education for Community Development: Aiming at Resilient Sustainable Society. In J. Vargas-Hernández & J. Zdunek-Wielgołaska (Eds.), *Bioeconomical Solutions and Investments in Sustainable City Development* (pp. 224–248). Hershey, PA: IGI Global. doi:10.4018/978-1-5225-7958-8.ch010

Moreira, A., & Ferreira, M. A. (2017). Strategic Challenges of the Portuguese Molds Industry: A Sectoral Innovation Perspective. In L. Carvalho (Ed.), *Handbook of Research on Entrepreneurial Development and Innovation Within Smart Cities* (pp. 534–560). Hershey, PA: IGI Global. doi:10.4018/978-1-5225-1978-2.ch023

Mujere, N., & Moyce, W. (2017). Climate Change Impacts on Surface Water Quality. In W. Ganpat & W. Isaac (Eds.), *Environmental Sustainability and Climate Change Adaptation Strategies* (pp. 322–340). Hershey, PA: IGI Global. doi:10.4018/978-1-5225-1607-1.ch012

Mundula, L., & Auci, S. (2017). Smartness, City Efficiency, and Entrepreneurship Milieu. In L. Carvalho (Ed.), *Handbook of Research on Entrepreneurial Development and Innovation Within Smart Cities* (pp. 173–198). Hershey, PA: IGI Global. doi:10.4018/978-1-5225-1978-2.ch009

Nath, R. (2019). Parametric Evaluation of Beam Deflection on Piezoelectric Material Using Implicit and Explicit Method Simulations: A Study in Energy Engineering. In J. Vargas-Hernández & J. Zdunek-Wielgołaska (Eds.), *Bioeconomical Solutions and Investments in Sustainable City Development* (pp. 65–87). Hershey, PA: IGI Global. doi:10.4018/978-1-5225-7958-8.ch004

Nayyar, A., Jain, R., Mahapatra, B., & Singh, A. (2019). Cyber Security Challenges for Smart Cities. In K. Ahuja & A. Khosla (Eds.), *Driving the Development, Management, and Sustainability of Cognitive Cities* (pp. 27–54). Hershey, PA: IGI Global. doi:10.4018/978-1-5225-8085-0.ch002

Odabasi, A., & Tiryaki, C. S. (2016). An Empirical Review of Long Term Electricity Demand Forecasts for Turkey. In U. Akkucuk (Ed.), *Handbook of Research on Waste Management Techniques for Sustainability* (pp. 227–243). Hershey, PA: IGI Global. doi:10.4018/978-1-4666-9723-2.ch012

Okay, E. (2016). Towards Smart Cities in Turkey?: Transitioning from Waste to Creative, Clean and Cheap Eco-Energy. In U. Akkucuk (Ed.), *Handbook of Research on Waste Management Techniques for Sustainability* (pp. 277–302). Hershey, PA: IGI Global. doi:10.4018/978-1-4666-9723-2.ch015

Pawlikowska-Piechotka, A., Łukasik, N., Ostrowska-Tryzno, A., & Sawicka, K. (2017). A Smart City Initiative: Urban Greens and Evaluation Method of the Sport and Recreation Potentials (SEM). In L. Carvalho (Ed.), *Handbook of Research on Entrepreneurial Development and Innovation Within Smart Cities* (pp. 561–583). Hershey, PA: IGI Global. doi:10.4018/978-1-5225-1978-2.ch024

Raisinghani, M. S., & Idemudia, E. C. (2016). Green Information Systems for Sustainability. In U. Akkucuk (Ed.), *Handbook of Research on Waste Management Techniques for Sustainability* (pp. 212–226). Hershey, PA: IGI Global. doi:10.4018/978-1-4666-9723-2.ch011

Rathee, D. S., Ahuja, K., & Hailu, T. (2019). Role of Electronics Devices for E-Health in Smart Cities. In K. Ahuja & A. Khosla (Eds.), *Driving the Development, Management, and Sustainability of Cognitive Cities* (pp. 212–233). Hershey, PA: IGI Global. doi:10.4018/978-1-5225-8085-0.ch009

Rotondo, F. (2018). The Grid Cities: Between Tradition and Innovation. In G. Carlone, N. Martinelli, & F. Rotondo (Eds.), *Designing Grid Cities for Optimized Urban Development and Planning* (pp. 109–127). Hershey, PA: IGI Global. doi:10.4018/978-1-5225-3613-0.ch007

Sánchez-Fernández, M. D., & Cardona, J. R. (2017). The Perception of the Effect of Tourism on the Local Community before the Ibiza Smart Island Project. In L. Carvalho (Ed.), *Handbook of Research on Entrepreneurial Development and Innovation Within Smart Cities* (pp. 392–416). Hershey, PA: IGI Global. doi:10.4018/978-1-5225-1978-2.ch017

Santos, B. (2017). Improving Urban Planning Information, Transparency and Participation in Public Administrations. *International Journal of E-Planning Research, 6*(4), 58–75. doi:10.4018/IJEPR.2017100104

Sardà, J. F. (2018). Cerdà/Barcelona/Eixample: 1855-2017 … A Work in Progress. In G. Carlone, N. Martinelli, & F. Rotondo (Eds.), *Designing Grid Cities for Optimized Urban Development and Planning* (pp. 19–38). Hershey, PA: IGI Global. doi:10.4018/978-1-5225-3613-0.ch002

Selvi, M. S. (2016). Physical Distribution Problems of Textile Companies in Turkey. In U. Akkucuk (Ed.), *Handbook of Research on Waste Management Techniques for Sustainability* (pp. 328–346). Hershey, PA: IGI Global. doi:10.4018/978-1-4666-9723-2.ch017

Signorile, N. (2018). Pride and Prejudice: The Murattiano-Modernism. In G. Carlone, N. Martinelli, & F. Rotondo (Eds.), *Designing Grid Cities for Optimized Urban Development and Planning* (pp. 52–77). Hershey, PA: IGI Global. doi:10.4018/978-1-5225-3613-0.ch004

Singh, G., Kapoor, R., & Khosla, A. K. (2019). Intelligent Anomaly Detection Video Surveillance Systems for Smart Cities. In K. Ahuja & A. Khosla (Eds.), *Driving the Development, Management, and Sustainability of Cognitive Cities* (pp. 163–182). Hershey, PA: IGI Global. doi:10.4018/978-1-5225-8085-0.ch007

Singh-Ackbarali, D., & Maharaj, R. (2017). Mini Livestock Ranching: Solution to Reducing the Carbon Footprint and Negative Environmental Impacts of Agriculture. In W. Ganpat & W. Isaac (Eds.), *Environmental Sustainability and Climate Change Adaptation Strategies* (pp. 188–212). Hershey, PA: IGI Global. doi:10.4018/978-1-5225-1607-1.ch007

Sood, N., Saini, I., Awasthi, T., Saini, M. K., Bhoriwal, P., & Kaur, T. (2019). Fog Removal Algorithms for Real-Time Video Footage in Smart Cities for Safe Driving. In K. Ahuja & A. Khosla (Eds.), *Driving the Development, Management, and Sustainability of Cognitive Cities* (pp. 55–86). Hershey, PA: IGI Global. doi:10.4018/978-1-5225-8085-0.ch003

Stone, R. J. (2017). Modelling the Frequency of Tropical Cyclones in the Lower Caribbean Region. In W. Ganpat & W. Isaac (Eds.), *Environmental Sustainability and Climate Change Adaptation Strategies* (pp. 341–349). Hershey, PA: IGI Global. doi:10.4018/978-1-5225-1607-1.ch013

Stratigea, A., Leka, A., & Panagiotopoulou, M. (2017). In Search of Indicators for Assessing Smart and Sustainable Cities and Communities' Performance. *International Journal of E-Planning Research, 6*(1), 43–73. doi:10.4018/IJEPR.2017010103

Touq, A. B., & Ijeh, A. (2018). Information Security and Ecosystems in Smart Cities: The Case of Dubai. *International Journal of Information Systems and Social Change, 9*(2), 28–43. doi:10.4018/IJISSC.2018040103

Üç, M., & Elitaş, C. (2016). Life Cycle Costing for Sustainability. In U. Akkucuk (Ed.), *Handbook of Research on Waste Management Techniques for Sustainability* (pp. 96–107). Hershey, PA: IGI Global. doi:10.4018/978-1-4666-9723-2.ch005

Ulker-Demirel, E., & Demirel, E. (2016). Green Marketing and Stakeholder Perceptions. In U. Akkucuk (Ed.), *Handbook of Research on Waste Management Techniques for Sustainability* (pp. 75–95). Hershey, PA: IGI Global. doi:10.4018/978-1-4666-9723-2.ch004

Vargas-Hernández, J. G., López, J. J., & Zdunek-Wielgołaska, J. A. (2019). Entrepreneurial and Institutional Analysis of Biodiesel Companies in Mexico. In J. Vargas-Hernández & J. Zdunek-Wielgołaska (Eds.), *Bioeconomical Solutions and Investments in Sustainable City Development* (pp. 89–115). Hershey, PA: IGI Global. doi:10.4018/978-1-5225-7958-8.ch005

Vázquez, D. G., & Gil, M. T. (2017). Sustainability in Smart Cities: The Case of Vitoria-Gasteiz (Spain) – A Commitment to a New Urban Paradigm. In L. Carvalho (Ed.), *Handbook of Research on Entrepreneurial Development and Innovation Within Smart Cities* (pp. 248–268). Hershey, PA: IGI Global. doi:10.4018/978-1-5225-1978-2.ch012

Velozo, R. A., & Montanha, G. K. (2017). Evaluation of a Mobile Software Development Company. In L. Carvalho (Ed.), *Handbook of Research on Entrepreneurial Development and Innovation Within Smart Cities* (pp. 514–533). Hershey, PA: IGI Global. doi:10.4018/978-1-5225-1978-2.ch022

About the Author

Timothy Ganesan is currently a Senior Analyst at the Royal Bank of Canada specializing in computational intelligence and data analytics. He has experience working as a Principal Researcher for the Fuels and Combustion Section in the research and development arm of the Malaysian power producer - Tenaga Nasional Berhad (TNB). In addition to having degrees in Chemical Engineering and Computational Fluid Dynamics, he holds a Ph.D. in Process Optimization. His research interests include engineering/industrial optimization, multi-objective/multi-level programming, evolutionary algorithms, machine learning, chaos optimization, and swarm-based optimization.

Index

www.ingramcontent.com/pod-product-compliance
Lightning Source LLC
LaVergne TN
LVHW061535290926
849162LV00004B/801